Smart Image Sensors and Applications

OPTICAL SCIENCE AND ENGINEERING

Founding Editor
Brian J. Thompson
University of Rochester
Rochester, New York

Smart CMOS Image Sensors and Applications

Jun Ohta

CRC Press
Taylor & Francis Group
Boca Raton   London   New York

CRC Press is an imprint of the
Taylor & Francis Group, an **informa** business

CRC Press
Taylor & Francis Group
6000 Broken Sound Parkway NW, Suite 300
Boca Raton, FL 33487-2742

ISBN-13: 978-1-1387-4681-7 (Paperback)

Library of Congress Cataloging-in-Publication Data

Ohta, Jun.
   Smart CMOS image sensors and applications / Jun Ohta.
      p. cm. -- (Optical science and engineering ; 129)
   Includes bibliographical references and index.
   ISBN 978-0-8493-3681-2 (hardback : alk. paper)
    1. Metal oxide semiconductors, Complementary--Design and construction. 2. Image processing--Digital techniques. I. Title.

   TK7871.99.M44O35 2007
   621.36'7--dc22
                                                   2007024953

Visit the Taylor & Francis Web site at
http://www.taylorandfrancis.com

and the CRC Press Web site at
http://www.crcpress.com

Preface

Image sensors have recently attracted renewed interest for use in digital cameras, mobile phone cameras, handy camcoders, cameras in automobiles, and other devices. For these applications, CMOS image sensors are widely used because they feature on-chip integration of the signal processing circuitry. CMOS image sensors for such specific purposes are sometimes called smart CMOS image sensors, vision chips, computational image sensors, etc.

Smart CMOS Image Sensors & Applications focuses on smart functions implemented in CMOS image sensors and their applications. Some sensors have already been commercialized, whereas some have only been proposed; the field of smart CMOS image sensors is active and generating new types of sensors. In this book I have endeavored to gather references related to smart CMOS image sensors and their applications; however, the field is so vast that it is likely that some topics are not described. Furthermore, the progress in the field is so rapid that some topics will develop as the book is being written. However, I believe the essentials of smart CMOS image sensors are sufficiently covered and that this book is therefore useful for graduate school students and engineers entering the field.

This book is organized as follows. First, MOS imagers and smart CMOS image sensors are introduced. The second chapter then describes the fundamental elements of CMOS image sensors and details the relevant optoelectronic device physics. Typical CMOS image sensor structures, such as the active pixel sensor (APS), are introduced in this chapter. The subsequent chapters form the main part of the book, namely a description of smart imagers. Chapter 3 introduces several functions for smart CMOS image sensors. Using these functions, Chapter 4 describes smart imaging, such as wide dynamic range image sensing, target tracking, and three-dimensional range finding. In the final chapter, Chapter 5, several examples of applications of smart CMOS image sensors are described.

This work is inspired by numerous preceding books related to CMOS image sensors. In particular, A. Moini's, "Vision Chips" [1], which features a comprehensive archive of vision chips, J. Nakamura's, "Image Sensors and Signal Processing for Digital Still Cameras" [2], which presents recent rich results of this field, K. Yonemoto's introductory but comprehensive book on CCD and CMOS imagers, "Fundamentals and Applications of CCD/CMOS Image Sensors" [3] and O. Yadid-Pecht and R. Etinne-Cummings's book, "CMOS Imagers: From Phototransduction To Image Processing" [4]. Of these, I was particularly impressed by K. Yonemoto's book, which unfortunately has only been published in Japanese. I hope that the present work helps to illuminate this field and that it complements that of Yonemoto's book.

I have also been influenced by books written by numerous other senior Japanese researchers in this field, including Y. Takamura [5], Y. Kiuchi [6] and T. Ando and H. Komobuchi [7]. The book on CCDs by A.J.P. Theuwissen is also useful [8].

I would like to thank the many people who have contributed both directly and indirectly to the areas covered in this book. Particularly, the colleagues in my laboratory, the Laboratory of Photonic Device Science in the Graduate School of Materials Science at the Nara Institute of Science and Technology (NAIST), Prof. Takashi Tokuda and Prof. Keiichiro Kagawa have made meaningful and significant contributions, which form the main parts of Chapter 5. This book would not have been born without their efforts. Prof. Masahiro Nunoshita for his continuous encouragement in the early stages of our laboratory. Kazumi Matsumoto, the secretary of our laboratory, for her constant support in numerous administrative affairs. Finally, the graduate students of my laboratory, both past and present, are thanked for their fruitful research. I would like to extend my most sincere thanks to all of these people.

For topics related to retinal prosthesis, I would like to thank Prof. Yasuo Tano, project leader of the Retinal Prosthesis Project, as well as Prof. Takashi Fujikado, Prof. Tetsuya Yagi, and Dr. Kazuaki Nakauchi at Osaka University. I would also like to thank the members of the Retinal Prosthesis Project at Nidek Co. Ltd., particularly Motoki Ozawa, Dr. Shigeru Nishimura, Kenzo Shodo, Yasuo Terasawa, Dr. Hiroyuki Kanda, Dr. Akihiro Uehara, and Naoko Tsunematsu. As advisory board members of the project, I would like to thank Prof. Emeritus Ryoichi Ito from the University of Tokyo and Prof. Emeritus Yoshiki Ichioka of Osaka University. The Retinal Prosthesis Project was supported by a Grant for Practical Application of Next-Generation Strategic Technology from the New Energy and Industrial Technology Development Organization (NEDO), Japan, and by Health and Labor Sciences Research Grants from the Ministry of Health, Labour, and Welfare of Japan. Of the members of the *in vivo* image sensor project, I would like to thank Prof. Sadao Shiosaka, Prof. Hideo Tamura, and Dr. David C. Ng. This work is partially supported by STARC (Semiconductor Technology Association Research Center). I also would like to thank Kunihiro Watanabe and colleagues for their collaborative research on demodulated CMOS image sensors.

I first entered the research area of smart CMOS image sensors as a visiting researcher at University of Colorado at Boulder under Prof. Kristina M. Johnson in 1992 to 1993. My experience there was very exciting and it helped me enormously in my initial research into smart CMOS image sensors after returning to Mitsubishi Electric Corporation. I would like to thank all of my colleagues at Mitsubishi Electric Corp. for their help and support, including Prof. Hiforumi Kimata, Dr. Shuichi Tai, Dr. Kazumasa Mitsunaga, Prof. Yutaka Arima, Prof. Masahiro Takahashi, Masaya Oita, Dr. Yoshikazu Nitta, Dr. Eiichi Funatsu, Dr. Kazunari Miyake, Takashi Toyoda, and numerous others.

I am most grateful to Prof. Masatoshi Ishikawa, Prof. Mitumasa Koyanagi, Prof. Jun Tanida, Prof. Shoji Kawahito, Prof. Richard Hornsey, Prof. Pamela Abshire, Yasuo Masaki, and Dr. Yusuke Oike for courteously allowing me to use their figures and data in this book. I have learned a lot from the committee members of the Institute of Image Information and Television Engineers (ITE), Japan, Prof. Masahide

Abe, Prof. Shoji Kawahito, Prof. Takayuki Hamamoto, Prof. Kazuaki Sawada, Prof. Junichi Akita, and the researchers of the numerous image sensor groups in Japan, including Dr. Shigeyuki Ochi, Dr. Yasuo Takemura, Takao Kuroda, Nobukazu Teranishi, Dr. Kazuya Yonemoto, Dr. Hirofumi Sumi, Dr. Yusuke Oike and others. I would particularly like to thank Taisuke Soda, who provided me with the opportunity to write this book, and Pat Roberson, both of Taylor & Francis/CRC, for their patience with me in the completion of this book. Without their continuous encouragement, completion of this book would not have been possible.

Personally, I extend my deepest thanks to Ichiro Murakami, for always stimulating my enthusiasm for image sensors and related topics. Finally, I would like to give special thanks to my wife Yasumi for her support and understanding during the long time it took to complete this book.

Jun Ohta,
Nara, July 2007

About the author

Jun Ohta was born in Gifu, Japan in 1958. He received his B.E., M.E., and Dr. Eng. degrees in applied physics, all from the University of Tokyo, Japan, in 1981, 1983, and 1992, respectively. In 1983, he joined Mitsubishi Electric Corporation, Hyogo, Japan, where he has been engaged in the research on optoelectronic integrated circuits, optical neural networks, and artificial retina chips. From 1992 to 1993, he was a visiting researcher in Optoelectronic Computing Systems Center, University of Colorado in Boulder. In 1998, he became an Associate Professor at the Graduate School of Materials Science, Nara Institute of Science and Technology (NAIST), in Nara, Japan, and in 2004, he became a Professor of NAIST. His current research interests include vision chips, CMOS image sensors, retinal prosthesis devices, bio-photonic LSIs, integrated photonic devices. Dr. Ohta received the Best Paper Award of the IEICE Japan in 1992, the Ichimura Award in 1996, the National Commendation for Invention in 2001, and the Niwa Takayanagi Award in 2007. He is a member of the Japan Society of Applied Physics, the Institute of Electronics, Information and Communication Engineers of Japan, the Institute of Image Information and Television Engineers of Japan, Japanese Society for Medical and Biological Engineering, the Institute of Electronic and Electronics Engineers, and the Optical Society of America.

Contents

1

Introduction

1.1 A general overview

Complementary metal–oxide–semiconductor (CMOS) image sensors have been the subject of extensive development and now share the market with charge coupled device (CCD) image sensors, which have dominated the field of imaging sensors for a long time. CMOS image sensors are now widely used not only for consumer electronics, such as compact digital still cameras (DSC), mobile phone cameras, handy-camcorders, and digital single lens reflex (DSLR) cameras, but also for cameras used for automobiles, surveillance, security, robot vision, etc. Recently, further applications of CMOS image sensors in biotechnology and medicine have emerged. Many of these applications require advanced performance such as wide dynamic range, high speed, and high sensitivity, while others need dedicated functions, such as real time target tracking and three-dimensional range finding. It is difficult to perform such tasks with conventional image sensors. Furthermore, some signal processing devices are insufficient for these purposes. Smart CMOS image sensors, CMOS image sensors with integrated smart functions on the chip, may meet the requirements of these applications.

CMOS image sensors are fabricated based on standard CMOS large scale integration (LSI) fabrication processes, while CCD image sensors are based on a specially developed fabrication process. This feature of CMOS image sensors makes it possible to integrate functional circuits to develop smart CMOS image sensors and to realize both a higher performance than that of CCDs and conventional CMOS image sensors and versatile functions that cannot be achieved with conventional image sensors.

Smart CMOS image sensors are mainly aimed at two different fields. One is the enhancement or improvement of the fundamental characteristics of CMOS image sensors, such as dynamic range, speed, and sensitivity. Another is the implementation of new functions, such as three-dimensional range finding, target tracing, and modulated light detection. For both fields, many architectures and/or structures, as well as materials, have been proposed and demonstrated.

The following terms are also associated with smart CMOS image sensors: computational CMOS image sensors, integrated functional CMOS image sensors, vision chips, focal plane image processing, as well as many others. Besides vision chips, these terms suggest that an image sensor has other functions in addition to imaging.

The name vision chip originates from a device proposed and developed by C. Mead and coworkers, which mimics the human visual system. It will be described later in this chapter. In the following section, we survey the history of CMOS image sensors and then review the brief history of smart CMOS image sensors.

1.2 Brief history of CMOS image sensors

The birth of MOS imagers

The history of MOS image sensors, shown in Fig. 1.1, starts with solid-sate imagers used as a replacement for image tubes. For solid-state image sensors, four important functions had to be realized: light-detection, accumulation of photo-generated signals, switching from accumulation to readout, and scanning. These functions are discussed in Chapter 2. The scanning function in X–Y addressed silicon-junction photosensing devices was proposed in the early 1960s by S.R. Morrison at Honeywell as the "photoscanner" [9] and by J.W. Horton et al. at IBM as the "scanistor" [10]. P.K. Weimer et al. proposed solid-state image sensors with scanning circuits using thin-film transistors (TFTs) [11]. In these devices, photoconductive film, discussed in Sec. 2.3.5, is used for the photodetector. M.A. Schuster and G. Strull at NASA used phototransistors (PTrs), which are discussed in Section 2.3.3, as photodetectors, as well as switching devices to realize X–Y addressing [12]. They successfully obtained images with a fabricated 50×50-pixel array sensor.

The accumulation mode in a photodiode is an important function for MOS image sensors and is described in Section 2.4. It was first proposed by G.P. Weckler at Fairchild Semiconductor [13]. In the proposal, the floating source of a MOSFET is used as a photodiode. This structure is used in some present CMOS image sensors. Weckler later fabricated and demonstrated a 100×100-pixel image sensor using this structure [14]. Since then, several types of solid-sate image sensors have been proposed and developed [14–17], as summarized in Ref. [18].

The solid-state image sensor developed by P.J. Noble at Plessey was almost the same as a MOS image sensor or passive pixel sensor (PPS), discussed in Section 2.5.1, consisting of a photodiode and a switching MOS transistor in a pixel with X- and Y-scanners and a charge amplifier. Noble briefly discussed the possibility of integrating logic circuitry for pattern recognition on a chip, which may be the first prediction of a smart CMOS image sensor.

Competition with CCDs

Shortly after the publication of details of solid-state image sensors in *IEEE Transaction on Electron Devices* in 1968, CCD image sensors appeared [19]. The CCD itself was invented in 1969 by W. Boyle and G.E. Smith at AT&T Bell Laboratories [19] and was experimentally verified at almost the same time [20]. Initially, the CDD was

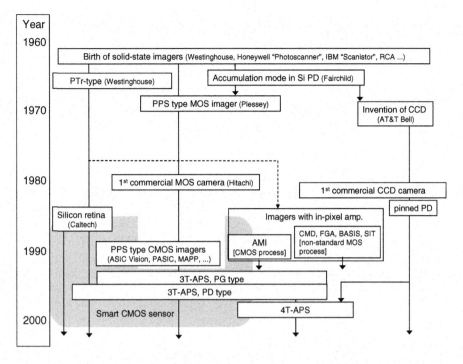

FIGURE 1.1

Evolution of MOS image sensors and related matters.

developed as semiconductor bubble memory, as a replacement for magnetic bubble memory, but was soon developed for use in image sensors. The early stages of the invention of the CCD is described in Ref. [21].

Considerable research effort resulted in the production of the first commercial MOS imagers, appearing in the 1980s [22–27]. While Hitachi and Matsushita have developed MOS imagers [28], until recently CCDs have been widely manufactured and used due to the fact that they have offered superior image quality than MOS imagers.

Solid-state imagers with in-pixel amplification

Subsequently, effort has been made to improve the signal-to-noise ratio (SNR) of MOS imagers by incorporating an amplification mechanism in a pixel. In the 1960s, a phototransistor (PTr) type imager was developed [12]. In the late 1980s, several amplified type imagers were developed, including the charge modulated device (CMD) [29], floating gate array (FGA) [30], base-stored image sensor (BASIS) [31], static induction transistor (SIT) type [32], amplified MOS imager (AMI) [33–37], and others [6, 7]. Apart from AMI, these required some modification of standard

MOS fabrication technology in the pixel structure, and ultimately they were not commercialized and their development was terminated. AMI can be fabricated in standard CMOS technology without any modification, however, and its pixel structure is the same as that of the active pixel sensor (APS); although AMI uses an I–V converter as a readout circuit while APS uses a source follower, though this difference is not critical. APS is also classified as an image sensor with in-pixel amplification.

Present CMOS image sensors

APS was first realized by using a photogate (PG) as a photodetector by E. Fossum et al. at JPL[*] and then by using a photodiode (PD) [38, 39]. A PG was used mainly due to ease of signal charge handling. The sensitivity of a PG is not as good since polysilicon as a gate material is opaque at the visible wavelength region. APSs using a PD are called 3T-APSs (three transistor APSs) and are now widely used in CMOS image sensors. In the first stage of 3T-APS development, the image quality could not compete with that of CCDs, both with respect to fixed pattern noise (FPN) and random noise. Introducing noise canceling circuits reduces FPN but not random noise.

By incorporating a pinned PD structure used in CCDs, which has a low dark current and complete depletion structure, the 4T-APS (four transistor APS) has been successfully developed [40]. A 4T-APS can be used with correlated double sampling (CDS), which can eliminate $k_B TC$ noise, the main factor in random noise. The image quality of 4T-APSs can compete with that of CCDs. The final issue for 4T-APSs is the large pixel size compared with that in CCDs. A 4T-APS has four transistors plus a PD and floating diffusion (FD) in a pixel, while a CCD has one transfer gate plus a PD. Although CMOS fabrication technology advances have benefited the development of CMOS image sensors [41], namely in shrinking the pixel size, it is essentially difficult to realize a smaller pixel size than that of CCDs. Recently, a pixel sharing technique has been widely used in 4T-APSs and has been effective in reducing the pixel size to be comparable with that of CCDs. Figure 1.2 shows the trend of pixel pitch in 4T-APSs. The figure illustrates that the pixel pitch of CMOS image sensors is comparable with that of CCDs, shown as open squares in the figure.

---

[*] Jet Propulsion Laboratory.

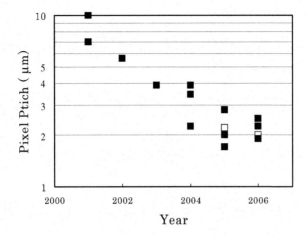

FIGURE 1.2

Trend of pixel pitch in 4T-APS type CMOS imagers. The solid and open squares show the pixel pitch for CMOS imagers and CCDs, respectively.

1.3 Brief history of smart CMOS image sensors

Vision chips

There are three main categories for smart CMOS image sensors, as shown in Fig. 1.3: pixel-level processing, chip-level processing or camera-on-a-chip, and column-level processing. The first category is vision chips or pixel-parallel processing. In the 1980s, C. Mead and coworkers at Caltech [*] proposed and demonstrated vision chips or silicon retina [42]. A silicon retina mimics the human visual processing system with massively parallel-processing capability using Si LSI technology. The circuits work in the subthreshold region, as discussed in Appendix E, in order to achieve low power consumption. In addition, the circuits automatically execute to solve a given problem by using convergence in two-dimensional resistive networks, as described in Section 3.3.3. They frequently use phototransistors (PTrs) as photodetectors due to the gain of PTrs. Since the 1980s, considerable work has been done on developing visions chips and similar devices, as reviewed by Koch and Liu [43], and A. Moini [1]. Massively parallel processing in the focal plane is very attractive and has been the subject of much research into fields such as programmable artificial retinas [44]. Some applications have been commercialized, such as two-layered resistive networks using 3T-APS by T. Yagi, S. Kameda, and co-workers at Osaka Univ. [45, 46].

---

[*]California Institute of Technology.

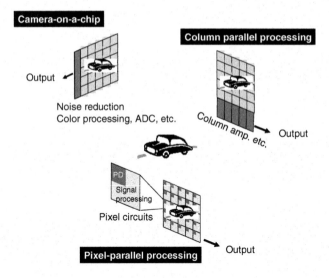

FIGURE 1.3
Three types of smart sensors.

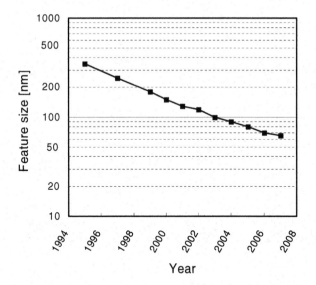

FIGURE 1.4
ITRS roadmap; the trend of DRAM half pitch [47].

Figure 1.4 shows the LSI roadmap of ITRS* [47]. This figure shows the trend

---

*International Technology Roadmap for Semiconductors.

of dynamic random access memory (DRAM) half pitch; other technologies such as logic processes exhibit almost the same trend, namely the integration density of LSI increases according to Moore's law such that the integration density doubles every 18—24 months [48].

This advancement of CMOS technology means that massively parallel processing or pixel-parallel processing is becoming more feasible. Considerable research has been published, such as vision chips based on cellular neural networks (CNN) [49–52], programmable multiple instruction and multiple data (MIMD) vision chips [53], biomorphic digital vision chips [54], and analog vision chips [55, 56]. Other pioneering work include digital vision chips using pixel-level processing based on a single instruction and multiple data (SIMD) processor by M. Ishikawa *et al.* at Univ. Tokyo and Hamamatsu Photonics [57–63].

It is noted that some vision chips are not based on the human visual processing system, and thus they belong in the category of pixel-level processing.

Advancement of CMOS technology and smart CMOS image sensors

The second category, pixel-level programming, is more tightly related with the advancement of CMOS technology and has little relation with pixel-parallel processing. This category includes system-on-chip and system-on-a-camera. In the early 1990s, advancement of CMOS technology made it possible to realize highly integrated CMOS image sensors or smart CMOS image sensors for machine vision. Pioneering work include ASIC vision (originally developed in Univ. Edinburgh [64,65] and later by VLSI Vision Ltd. (VVL)), near-sensor image processing (NSIP) (later known as PASIC [66]) originally developed in Linköping University [67] and MAPP by Integrated Vision Products (IVP) [68].

PASIC may be the first CMOS image sensor that uses a column level analog-to-digital converter (ADC) [66]. ASIC vision has a PPS structure [67], while NSIP uses a pulse width modulation (PWM) based sensor [67], discussed in Section 3.4.1. MAPP uses an APS [68].

Smart CMOS image sensors based on high performance CMOS image sensor technologies

Some of the above mentioned sensors have column-parallel processing structure, the third of the categories. Column-parallel processing is suitable for CMOS image sensors, because the column lines are electrically independent of each other. Column-parallel processing can enhance the performance of CMOS image sensors, such as widening the dynamic range and increasing the speed. Combining with 4T-APSs, column-parallel processing exhibit a high image quality and versatile functions. Therefore, recently column-level processing architecture has been widely used for higher performance CMOS image sensors. The advancement of LSI technologies also broadens the range of applications of this architecture.

1.4   Organization of the book

This book is organized as follows. First, in this introduction, a general overview of solid-state image sensors is presented. Then, smart CMOS image sensors are described including a brief history and a discussion of their features. Next, in Chapter 2, fundamental information on CMOS image sensors is presented in detail. First, optoelectronic properties of silicon semiconductors, based on CMOS technology, are described in Section 2.2. Then, in Section 2.3, several types of photodetectors are introduced, including the photodiode, which is commonly used in CMOS image sensors. The operation principle and fundamental characteristics of photodiodes are described. In a CMOS image sensor, a photodiode is used in the accumulation mode, which is very different from the mode of operation for other applications such as optical communication. The accumulation mode is discussed in Section 2.4. Pixel structure is the heart of this chapter and is explained in Section 2.5 for active pixel sensors (APS) and passive pixel sensors (PPS). Peripheral blocks other than pixels are described in Section 2.6. Addressing and readout circuits are also mentioned in that section. The fundamental characteristics of CMOS image sensors are discussed in Section 2.7. The topics of color (Section 5.4.1) and pixel sharing (Section 2.9) are also described in this chapter. Finally, several comparisons are discussed in Sections 2.10 and 2.11.

In Chapter 3, several smart functions and materials are introduced. Certain smart CMOS image sensors have been developed by introducing new functions into conventional CMOS image sensor architecture. Firstly, pixel structures different from that of conventional APS are introduced in Section 3.2, such as the log sensor. Smart CMOS image sensors can be classified into three categories, analog, digital, and pulse, described in Sections 3.3, 3.4, and 3.5. CMOS image sensors are typically based on silicon CMOS technologies, but other technologies and materials can be used to achieve smart functions. For example, silicon on sapphire (SOS) technology is a candidate for smart CMOS image sensors. Section 3.6 discussed materials other than silicon in smart CMOS image sensors. Structures other than standard CMOS technologies are described in Section 3.7.

By combining the smart functions introduced in Chapter 3, Chapter 4 describes several examples of smart imaging. Low-light imaging (Section 4.2), high speed (Section 4.3), and wide dynamic range (Section 4.4) are presented with examples. These features of smart CMOS image sensors give a higher performance compared with conventional CMOS image sensors. Another feature of smart CMOS image sensors is to achieve versatile functions that cannot be realized by conventional image sensors. For this, sections on demodulation (Section 4.5), three-dimensional range finders (Section 4.6), and target tracking (Section 4.7) are presented. Finally in this chapter, dedicated arrangements of pixels and optics are described. Section 4.8 considers two types of smart CMOS image sensors with nonorthogonal pixel arrangements and dedicated optics.

The final chapter, Chapter 5, considers applications using smart CMOS image sensors in the field of information and communication technologies, biotechnologies, and medicine. These applications have recently emerged and will be important for the next generation of smart CMOS image sensors.

Several appendices are attached to present additional information for the main body of the book.

2

Fundamentals of CMOS image sensors

2.1 Introduction

This chapter provides the fundamental knowledge for understanding CMOS image sensors.

A CMOS image sensor generally consists of an imaging area, which consists of an array of pixels, vertical and horizontal access circuitry, and readout circuitry, as shown in Fig. 2.1.

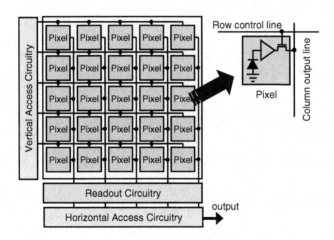

FIGURE 2.1

Architecture of a CMOS image sensor. A two-dimensional array of pixels, vertical and horizontal access circuitry, and readout circuitry are generally implemented. A pixel consists of a photodetector and transistors.

The imaging area is a two-dimensional array of pixels; each pixel contains a photodetector and some transistors. This area is the heart of an image sensor and the imaging quality is largely determined by the performance of this area. Access circuitry is used to access a pixel and read the signal value in the pixel. Usually a scanner or shift register is used for the purpose, and a decoder is used to access pixels randomly, which is sometimes important for smart sensors. A readout circuit is a one-dimensional array of switches and a sample and hold (S/H) circuit. Noise cancel circuits, such as correlated double sampling (CDS), are employed in this area.

In this chapter, these fundamental elements of CMOS image sensors are described. First, photodetection is explained. The behavior of minority carriers plays an important role in photodetection. Several kinds of photodetectors for CMOS image sensors have been introduced. Among them, pn-junction photodiodes are most often used, and the operation principle and hence the basic characteristics of pn-junction photodiodes are explained here in detail. In addition, the accumulation mode, which is an important operation for CMOS image sensors, is described. Then, basic pixel structures are introduced, namely passive pixel sensors and active pixel sensors. Finally, further elements for CMOS image sensors are described, such as scanners and decoders, read-out circuits, and noise cancellers.

2.2   Fundamentals of photodetection

2.2.1   Absorption coefficient

When light is incident on a semiconductor, a part of the incident light is reflected while the rest is absorbed in the semiconductor and produces electron–hole pairs inside the semiconductor, as shown in Fig. 2.2. Such electron–hole pairs are called photo-generated carriers. The amount of photo-generated carriers depends on the semiconductor material and is described by the absorption coefficient α.

It should be noted that α is defined as the ratio of decrease of light power $\Delta P/P$ when the light travels a distance Δz, that is,

$$\alpha(\lambda) = \frac{1}{\Delta z}\frac{\Delta P}{P}. \tag{2.1}$$

From Eq. 2.1, the following relation is derived:

$$P(z) = P_o \exp(-\alpha z). \tag{2.2}$$

The absorption length L_{abs} is defined as

$$L_{abs} = \alpha^{-1}. \tag{2.3}$$

It is noted that the absorption coefficient is a function of photon energy $h\nu$ or wavelength λ, where h and ν are Planck's constant and the frequency of the light. The value of the absorption length L_{abs} thus depends on wavelength. Figure 2.3 shows the dependence of the absorption coefficient and the absorption length of silicon on the input light wavelength. In the visible region, 0.4–0.6 μm, the absorption length lies within 0.1–10 μm [69]. The absorption length is an important figure for a rough estimation of a photodiode structure.

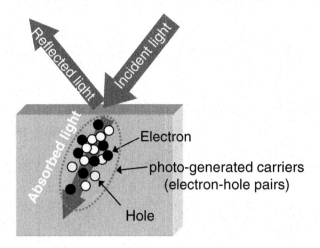

FIGURE 2.2
Photo-generated carriers in a semiconductor.

2.2.2 Behavior of minority carriers

Incident light on a semiconductor generates electron–hole pairs or photo-generated carriers. When electrons are generated in a p-type region, the electrons are minority carriers. The behavior of minority carriers is important for image sensors. For example, in a CMOS image sensor with a p-type substrate, photo-generated minority carriers in the substrate are electrons. This situation occurs when infrared (IR) light is incident in the sensor, because the absorption length in the IR region is over 10 μm, as shown in Fig. 2.3, and thus the light reaches the substrate. In that case, the diffusion behavior of the carriers greatly affects the image sensor characteristics; they can diffuse to adjacent photodiodes through the substrate and cause image blurring. To suppress this, an IR cut filter is usually used, because IR light reaches deeper regions of the photodiode, namely the substrate, and produces more carriers than visible light in the deeper regions.

The mobility and lifetime of minority carriers are empirically given by the following relations [70–72] with parameters of acceptor concentration N_a and donor concentration N_d:

$$\mu_n = 233 + \frac{1180}{1 + [N_a/(8 \times 10^{16})]^{0.9}} \ [\text{cm}^2/\text{V} \cdot \text{s}], \tag{2.4}$$

$$\mu_p = 130 + \frac{370}{1 + [N_d/(8 \times 10^{17})]^{1.25}} \ [\text{cm}^2/\text{V} \cdot \text{s}], \tag{2.5}$$

$$\tau_n^{-1} = 3.45 \times 10^{-12} N_a + 0.95 \times 10^{-31} N_a^2 \ [\text{s}^{-1}], \tag{2.6}$$

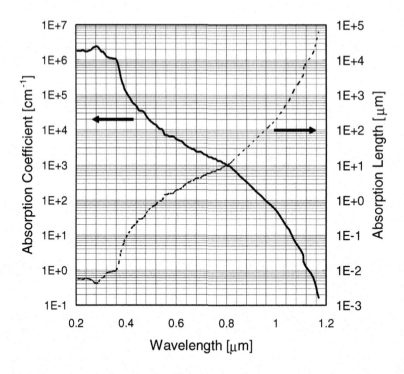

FIGURE 2.3
Absorption coefficient (solid line) and absorption length (broken line) of silicon as a
function of wavelength. From the data in [69].

$$\tau_p^{-1} = 7.8 \times 10^{-13} N_a + 1.8 \times 10^{-31} N_d^2 \ [\text{s}^{-1}]. \tag{2.7}$$

From the above equations, we can estimate the diffusion lengths $L_{n,p}$ for electrons
and holes by using the relation

$$L_{e,p} = \sqrt{\frac{k_B T \mu_{n,p} \tau_{n,p}}{e}}. \tag{2.8}$$

Figure 2.4 shows the diffusion lengths of electrons and holes as a function of
impurity concentration. Note that both electrons and holes can travel over 100 μm
for impurity concentrations below 10^{17} cm^{-3}.

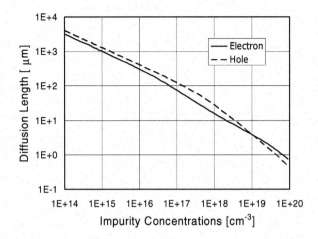

FIGURE 2.4
Diffusion lengths of electrons and holes in silicon as a function of impurity concentration.

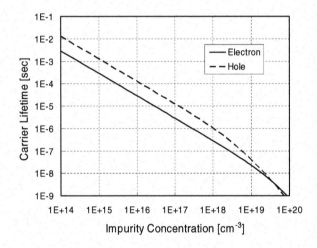

FIGURE 2.5
Lifetimes of electrons and holes in silicon as a function of impurity concentration.

2.2.3 Sensitivity and quantum efficiency

The sensitivity is defined as the amount of photocurrent I_L produced when a unit of light power P_o is incident on a material. It is given by

$$R_{ph} \equiv \frac{I_L}{P_o}. \tag{2.9}$$

The quantum efficiency is defined as the ratio of the number of generated photocarriers to the number of the input photons. The input photon number per unit time and the generated carrier number per unit time are I_L/e and $P_o/(h\nu)$, and thus the quantum efficiency is expressed as

$$\eta_Q \equiv \frac{I_L/e}{P_o/(h\nu)} = R_{ph}\frac{h\nu}{e}. \tag{2.10}$$

From Eq. 2.10, the maximum sensitivity, that is the sensitivity at $\eta_Q = 1$, is found to be

$$R_{ph,max} = \frac{e}{h\nu} = \frac{e}{hc}\lambda = \frac{\lambda\,[\mu m]}{1.23}. \tag{2.11}$$

$R_{ph,max}$ is illustrated in Fig. 2.6. It monotonically increases in proportion to the wavelength of the input light and eventually reaches zero at the wavelength λ_g corresponding to the bandgap of the material E_g. For silicon, the wavelength is about 1.12 μm since the bandgap of silicon is 1.107 eV.

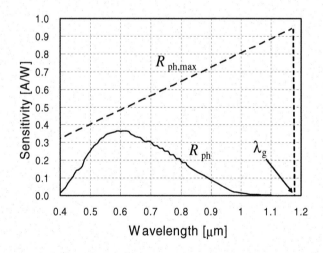

FIGURE 2.6
Sensitivity of silicon. The solid line shows the sensitivity R_{ph} according to Eq. 2.19. The dashed line shows the ideal sensitivity or maximum sensitivity $R_{ph,max}$ according to Eq. 2.11. λ_g is the wavelength at the bandgap of silicon.

2.3   Photodetectors for smart CMOS image sensors

Most photodetectors used in CMOS image sensors are pn-junction photodiodes (PDs).
In the next sections, PDs are described in detail. Other photodetectors used in CMOS
image sensors are photogates (PGs), phototransistors (PTrs), and avalanche photodiodes (APDs). PTrs and APDs both make use of gain; another detector with gain
is the photoconductive detector (PCD). Figure 2.7 illustrates the structures of PDs,
PGs, and PTrs.

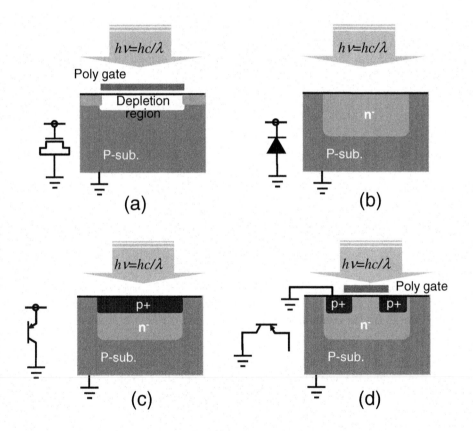

FIGURE 2.7
Symbols and structures of (a) photodiode, (b) photogate, (c) vertical type phototransistor, and (d) lateral type phototransistor.

2.3.1 pn-junction photodiode

In this section, a conventional photodetector, a pn-junction PD, is described [73, 74]. First, the operation principle of a PD is described and then several fundamental characteristics, such as quantum efficiency, sensitivity, dark current, noise, surface recombination, and speed, are discussed. These characteristics are important for smart CMOS image sensors.

2.3.1.1 Operation principle

The operation principle of the pn-junction PD is quite simple. In a pn-junction diode, the forward current I_F is expressed as

$$I_F = I_{diff}\left[\exp\left(\frac{eV}{nk_BT}\right) - 1\right], \tag{2.12}$$

where n is an ideal factor and I_{diff} is the saturation current or diffusion current, which is given by

$$I_{diff} = eA\left(\frac{D_n}{L_n}n_{po} + \frac{D_p}{L_p}p_{no}\right), \tag{2.13}$$

where $D_{n,p}$, $L_{n,p}$, n_{po}, and p_{no} are the diffusion coefficient, diffusion length, minority carrier concentration in the p-type region, and the minority carrier concentration in n-type region, respectively. A is the cross-section area of the pn-diode. The photocurrent of the pn-junction photodiode is expressed as follows:

$$\begin{aligned} I_L &= I_{ph} - I_F \\ &= I_{ph} - I_{diff}\left[\exp\left(\frac{eV}{nk_BT}\right) - 1\right], \end{aligned} \tag{2.14}$$

where n is an ideal factor. Figure 2.8 illustrates the I–V curves of a pn-PD under dark and illuminated conditions. There are three modes for bias conditions: solar cell mode, PD mode, and avalanche mode, as shown in Fig. 2.8.

Solar cell mode In the solar cell mode, no bias is applied to the PD. Under light illumination, the PD acts as a battery, that is it produces a voltage across the pn-junction. Figure 2.8 shows the open circuit voltage V_{oc}. In the open circuit condition, the voltage V_{oc} can be obtained from $I_L = 0$ in Eq. 2.14, and thus

$$V_{oc} = \frac{k_BT}{e}\ln\left(\frac{I_{ph}}{I_{diff}} + 1\right). \tag{2.15}$$

This shows that the open circuit voltage does not linearly increase according to the input light intensity.

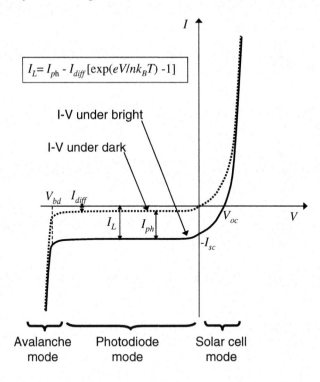

$$I_L = I_{ph} - I_{diff}[\exp(eV/nk_BT) - 1]$$

I-V under bright

I-V under dark

V_{bd} I_{diff}

I_L I_{ph}

V_{oc}

$-I_{sc}$

Avalanche mode Photodiode mode Solar cell mode

FIGURE 2.8
PD I–V curves under dark and bright conditions.

PD mode   The second mode is the PD mode. When a PD is reverse biased, that is $V < 0$, the exponential term in Eq. 2.14 can be neglected, and thus I_L becomes

$$I_L \approx I_{ph} + I_{diff}. \tag{2.16}$$

This shows that the output current of the PD is equal to the sum of the photocurrent and diffusion current. Thus, the photocurrent lineally increases according to the input light intensity.

Avalanche mode   The third mode is the avalanche mode. When a PD is strongly biased, the photocurrent suddenly increases, as shown in Fig. 2.8. This phenomena is called an avalanche, where impact ionization of electrons and holes occurs and the carriers are multiplied. The voltage where an avalanche occurs is called the avalanche breakdown voltage V_{bd}, shown in Fig. 2.8. Avalanche breakdown is explained in Sec. 2.3.1.3. The avalanche mode is used in an avalanche photodiode (APD) and is described in Sec. 2.3.4.

2.3.1.2   Quantum efficiency and sensitivity

By using the definition of the absorption coefficient in Eq. 2.2 $\alpha(\lambda)$, the light intensity is expressed as

$$dP(z) = -\alpha(\lambda)P_o \exp[-\alpha(\lambda)z]\,dz. \tag{2.17}$$

To make it clear that the absorption coefficient is dependent on the wavelength, α is written as $\alpha(\lambda)$. The quantum efficiency is defined as the ratio of absorbed light intensity to the total input light intensity, and thus

$$\eta_Q = \frac{\int_{x_n}^{x_p} \alpha(\lambda)P_o \exp[-\alpha(\lambda)x]\,dx}{\int_0^\infty \alpha(\lambda)P_o \exp[-\alpha(\lambda)x]\,dx}$$
$$= (1 - \exp[-\alpha(\lambda)W])\exp[-\alpha(\lambda)x_n], \tag{2.18}$$

where W is the depletion width and x_n is the distance from the surface to the edge of the depletion region as shown in Fig. 2.9.

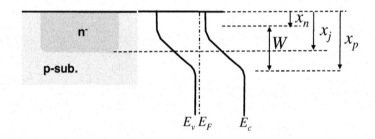

FIGURE 2.9

pn-junction structure. The junction is formed at a position x_j from the surface. The depletion region widens at the sides of the n-type region x_n and p-type region x_p. The width of the depletion region W is thus equal to $x_n - x_p$.

Using Eq. 2.18, the sensitivity is expressed as follows:

$$R_{ph} = \eta_Q \frac{e\lambda}{hc}$$
$$= \frac{e\lambda}{hc}(1 - \exp[-\alpha(\lambda)W])\exp[-\alpha(\lambda)x_n]. \tag{2.19}$$

In this equation, the depletion width W and the part of the depletion width at the N region x_n are expressed as follows. Using the built-in potential V_{bi}, W under an applied voltage V_{appl} is expressed as

$$W = \sqrt{\frac{2\varepsilon_{Si}(N_d + N_a)(V_{bi} + V_{appl})}{eN_aN_d}}, \tag{2.20}$$

where ε_{Si} is the dielectric constant of silicon. The built-in potential V_{bi} is given by

$$V_{bi} = k_B T \ln \left(\frac{N_d N_a}{n_i^2} \right), \tag{2.21}$$

where n_i is the intrinsic carrier concentration for silicon and $n_i = 1.4 \times 10^{10}$ cm^{-3}. The parts of the depletion width at the n-region and p-region are

$$x_n = \frac{N_a}{N_a + N_d} W, \tag{2.22}$$

$$x_p = \frac{N_d}{N_a + N_d} W. \tag{2.23}$$

Figure 2.6 shows the sensitivity spectrum curve of silicon, that is, the dependence of the sensitivity on the input light wavelength. The sensitivity spectrum curve is dependent on the impurity profile of the n-type and p-type regions as well as the position of the pn-junction x_j. In the calculation of the curve in Fig. 2.6, the impurity profile in both the n-type and p-type regions is flat and the junction is abrupt. In addition, only photo-generated carriers in the depletion region are accounted for; some portion of the photo-generated carriers outside the depletion region diffuse and reach the depletion region, but in the calculation these diffusion carriers are not accounted for. Such diffusion carriers can affect the sensitivity at long wavelengths because of the low value of the absorption coefficient in the long wavelength region [75]. Another assumption is to neglect the surface recombination effect, which will be considered in the section on noise, Sec. 2.3.1.4. These assumptions will be discussed in Sec. 2.4. An actual PD in an image sensor is coated with SiO$_2$ and Si$_3$N$_4$, and thus the quantum efficiency is changed [76].

2.3.1.3 Dark current

Dark current in PDs has several sources.

Diffusion current The diffusion current inherently flows and is expressed as

$$\begin{aligned}
I_{diff} &= Ae \left(\frac{D_n n_{po}}{L_n} + \frac{D_p n_{no}}{L_p} \right) \\
&= Ae \left(\frac{D_n}{L_n N_A} + \frac{D_p}{L_p N_D} \right) N_c N_v \exp \left(-\frac{E_g}{k_B T} \right),
\end{aligned} \tag{2.24}$$

where A is the diode area, N_c and N_v are the effective density of states in the conduction band and valence band, respectively, and E_g is the bandgap. Thus, the diffusion current exponentially increases as the temperature increases. It is noted that the diffusion current weakly depends on the bias voltage; more precisely, it depends on the square root of the bias voltage.

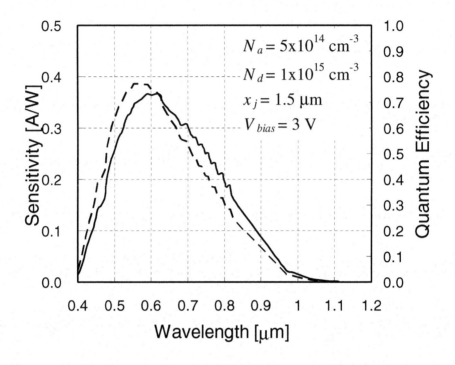

FIGURE 2.10
Dependence of the sensitivity (solid line) and quantum efficiency (broken line) of a
pn-junction PD on wavelength. The PD parameters are summarized in the inset.

Tunnel current   Other dark currents include tunnel current, generation–recombination
(g–r current), Frakel–Poole current, and surface leak current [77,78]. The tunnel current consists of band-to-band tunneling (BTBT) and trap-assisted tunneling (TAT),
which has an exponential dependence on the bias voltage [77–79] but has little dependence on the temperature. Although both BTBT and TAT cause dark current
exponentially dependent on bias voltage, the dependence is different, as shown in
Table 2.1. The tunnel current is important when doping is large and thus the depletion width becomes thin so as to lead to tunneling.

G–R current   In the depletion region, the carrier concentration is reduced and carrier generation occurs rather than recombination of carriers [77, 80] by thermal generation. This causes dark current. The g–r current is given by [77]

$$I_{gr} = AW \frac{e n_i}{\tau_g} = AW \frac{e \sqrt{N_c N_v}}{\tau_g} \exp\left(-\frac{E_g}{2 k_B T}\right), \tag{2.25}$$

where W is the depletion width, τ_g is the lifetime of the deep level, and n_i is the
intrinsic carrier concentration. This process is called Shockley–Read–Hall recombi-

nation [77, 80].

Impact ionization current Impact ionization or avalanche breakdown increases dark current when the bias voltage increases [81, 82]. The bias dependence of dark current by impact ionization arises from the voltage dependence of the ionization coefficients of electrons and holes, α_n and α_p. These coefficients exponentially increase by as the bias voltage increases.

Frankel–Poole current The Frankel–Poole current originates from the emission of trapped electrons into the conduction band [77]. This current strongly depends on the bias voltage, which is the same as the tunneling current.

Surface leak current The surface leak current is given by

$$I_{surf} = \frac{1}{2} e n_i s_o A_s, \tag{2.26}$$

where n_i, s_o, and A_s are the intrinsic carrier concentration, surface recombination rate, and surface area, respectively.

TABLE 2.1
Dependence of dark currents on temperature and voltage [77].
a, a', b, c are constants.

Process	Dependence
Diffusion	$\propto \exp\left(-\frac{E_g}{k_B T}\right)$
G–R	$\propto \sqrt{V} \exp\left(-\frac{E_g}{2k_B T}\right)$
Band-to-band tunneling	$\propto V^2 \exp\left(\frac{-a}{V}\right)$
Trap-assisted tunneling	$\propto \exp\left(\frac{-a'}{V}\right)^2$
Impact ionization	$\alpha \propto \exp\left(\frac{-b}{V}\right)$
Frankel–Poole	$\propto V \exp\left(\frac{-c}{T}\right)$
Surface leak	$\propto \exp\left(-\frac{E_g}{2k_B T}\right)$

Dependence of dark current on temperature and bias voltage   Comparing Eqs. 2.24, 2.25, and 2.26 shows that the temperature dependences of the various dark currents are different; only I_{surf} is independent of temperature, while $\log I_{diff}$ and $\log I_{gr}$ vary as $-\frac{1}{T}$ and $-\frac{1}{2T}$, respectively. Thus, the temperature dependence can reveal the origin of the dark current. Also, the dependence on the bias voltage is different. The dependence on temperature and bias voltage is summarized in Table 2.1.

2.3.1.4   Noise

Shot noise   A PD suffers from shot noise and thermal noise. Shot noise originates from fluctuations in the number of the particles N such as electrons and photons. Thus shot noise and electron (or hole) shot noise inherently exist in a PD. The root mean square of the shot noise current i_{sh} is expressed as

$$i_{sh,\,rms} = \sqrt{2e\bar{I}\Delta f}, \tag{2.27}$$

where $\bar{I}$ and Δf indicate average signal current and bandwidth, respectively. The signal-to-noise ratio (SNR) for shot noise is expressed as

$$\text{SNR} = \frac{\bar{I}}{\sqrt{2e\bar{I}\Delta f}} = \frac{\sqrt{I}}{2e\Delta f}. \tag{2.28}$$

Thus, as the amount of current or the number of electrons decreases, the SNR associated with shot noise decreases. Dark current also produces shot noise.

Thermal noise   In a load resistance R, free electrons exist and randomly move according to the temperature of the load resistance. This effect generates thermal noise, also known as Johnson noise or Nyquist noise. The thermal noise is expressed as

$$i_{sh,\,rms} = \sqrt{\frac{4k_B T \Delta f}{R}}. \tag{2.29}$$

In CMOS image sensors, the thermal noise appears as $k_B T C$ noise, which is discussed in Sec. 2.7.1.2.

2.3.1.5   Surface recombination

In a conventional CMOS image sensor, the surface of the silicon is interfaced with SiO_2 and has some dangling bonds, which produce surface states or interface states acting as non-recombination centers. Some photo-generated carriers near the surface are trapped at the centers and do not contribute to the photocurrent. Thus these surface states degrade the quantum efficiency or sensitivity. This effect is called surface recombination. The feature parameter for surface recombination is the surface recombination velocity S_{surf}. The surface recombination rate is proportional to the excess carrier density at the surface:

$$D_n \frac{\partial n_p}{\partial x} = S_{surf}\left[n_p(0) - n_{po}\right]. \tag{2.30}$$

The recombination velocity is strongly dependent on the interface state, band bending, defects, and other effects, and is approximately 10 cm^3/sec for both electrons and holes. For short wavelengths, such as blue light, the absorption coefficient is large and absorption mostly occurs at the surface. Thus it is important to reduce the surface recombination velocity to achieve high quantum efficiency in the short wavelength region.

2.3.1.6 Speed

In the recent growth of optical fiber communications and fiber-to-the-home (FTTH) technology, silicon CMOS photoreceivers have been extensively studied and developed. High-speed photodetectors using CMOS technologies, including BiCMOS technology, are described in detail in Ref. [83, 84], and high-speed circuits for CMOS optical fiber communication are also detailed in Ref. [85].

In conventional image sensors, the speed of the PD is not a concern. However, some kinds of smart image sensors need a PD with a fast response. Smart CMOS sensors for optical wireless LANs is an example and is considered in Chapter 5; they are based on technologies for CMOS-based photoreceivers for optical fiber communications, mentioned above. Another example is a smart CMOS sensor that can measure time-of-flight (TOF), also described in Chapter 5. In this case, an APD is used for its high-speed response.

Here we consider the response of a PD. Generally, the response of a PD is limited by the CR time constant τ_{CR}, transit time τ_{tr}, and diffusion time of minority carriers τ_n for electrons:

- The CR time originates from the pn-junction capacitance C_D and is expressed as

$$\tau_{CR} = 2\pi C_D R_L, \tag{2.31}$$

  where R_L is the load resistance.

- The transit time is defined as the time for a carrier to drift across the depletion region. It is expressed as

$$\tau_{tr} = W/v_s, \tag{2.32}$$

  where v_s is the saturation velocity.

- Minority carriers generated outside the depletion region can reach the depletion region after the diffusion time,

$$\tau_{n,p} = L_{n,p}^2/D_{n,p}, \tag{2.33}$$

  for electrons with a diffusion coefficient of D_n.

It is noted there is a trade-off between depletion width W and quantum efficiency η_Q in the case of transit time limitation. In this case,

$$\eta_Q = [1 - \exp(-\alpha(\lambda)v_s t_{tr})]\exp(-\alpha(\lambda)x_n). \tag{2.34}$$

Of these, the diffusion time has the greatest effect on the PD response in CMOS image sensors.

2.3.2   Photogate

The structure of a photogate (PG) is the same as a MOS capacitor; photo-generated carries accumulate in the depletion region when the gate is biased. A PG has a suitable structure to accumulate and transfer carriers, and PGs have been used in some CMOS image sensors. The accumulation of photo-generated carriers in a PG is shown in Fig. 2.11. By applying a gate bias voltage, a depletion region is produced and acts as an accumulation region for photo-generated carriers, as shown in Fig. 2.11.

The fact that the photo-generated area is separated from the top surface in a PG is useful for some smart CMOS image sensors, as will be discussed in Chapter 5. It is noted that PGs have disadvantages with regard to sensitivity, because the gate, which is usually made of polysilicon, is partially transparent and has an especially low transmittance at shorter wavelength or in the blue light region.

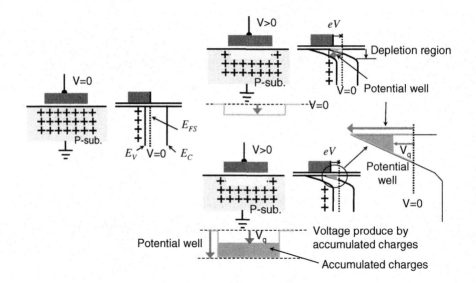

FIGURE 2.11
Photogate structure with applied gate voltage, which produces a depletion region where photo-generated carriers accumulate.

2.3.3   Phototransistor

A phototransistor (PTr) can be made using standard CMOS technology for a parasitic transistor. A PTr amplifies a photocurrent by a factor of the base current gain β. Because the base width and carrier concentration are not optimized by standard

CMOS process technology, β is not high, typically about 10–20. In particular, the base width is a trade-off factor for a phototransistor; when the base width increases, the quantum efficiency increases but the gain decreases [86]. Another disadvantage of a PTr is the large variation of β, which produces a fixed pattern noise (FPN), as detailed in Sec. 2.7.1.1. In spite of these disadvantages, PTrs are used in some CMOS image sensors due to their simple structure and their gain. When accompanied by current mirror circuits, PTrs can be used in current-mode signal processing, as discussed in Sec. 3.2.1. To address the low β at low photocurrent, a vertical inversion-layer emitter pnp BJT structure has been developed [87].

2.3.4 Avalanche photodiode

An avalanche photodiode (APD) utilizes an avalanche effect in which photo-generated carriers are multiplied [86]. APDs have a gain as well as a high-speed response. APDs are thus used as detectors in optical fiber communication and ultra low light detection such as biotechnologies. However, they are hardly used in image sensors, because they require a high voltage over 100 V. Such a high voltage hinders the use of APDs in standard CMOS technologies besides hybrid image sensors with other APD materials with a CMOS readout circuit substrate, as reported in Ref. [88] for example. Gain variation causes the same problem as seen in PTrs.

Pioneering work by A. Biber et al. at Centre Suisse d'Electronique et de Microtechnique (CSEM) has produced a 12 × 24-pixel APD array fabricated in standard 1.2-μm BiCMOS technology [89]. Each pixel employs an APD control and readout circuits. An image is obtained with the fabricated sensor with an avalanche gain of about 7 under a bias voltage of 19.1 V.

Several reports have been published of APDs fabricated using standard CMOS technologies [90–100], as shown in Fig. 2.12. In these reports, the APD is biased over the avalanche breakdown voltage, and thus when photons are incident on the APD, it quickly turns on and produces a spike-like current pulse. This phenomenon resembles that of a Geiger counter and thus it is called the Geiger mode. The Geiger mode is difficult to use in imaging, though it can be used in another applications, as described in Chapter 5.

Recently, H. Finkelstein et al. at Univ. California, San Diego have reported a Geiger-mode APD fabricated in 0.18-μm CMOS technology [101]. They use shallow trench isolation (STI) as a guard ring for the APD. A bias voltage of 2.5 V is found to be sufficient to achieve avalanche breakdown. This result suggests that deep sub-micron technology can be used to fabricate a CMOS image sensor with a single photon avalanche diode (SPAD) pixel array.

2.3.5 Photoconductive detector

Another gain detector is the photoconductive detector (PCD), which uses the effect of photoconductivity [86]. A PCD typically has a structure of n^+–n^-–n^+. A DC bias is applied between the two n^+ sides, and thus the generated electric field is largely confined to the n^- region, which is a photoconductive region where electron–hole

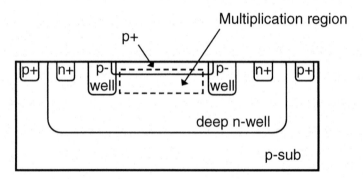

FIGURE 2.12
Avalanche photodiode structure using standard CMOS technology [98].

pairs are generated. The gain originates from a large ratio of the long lifetime of holes τ_p to the short transit time of electrons t_{tr}, that is $\tau_p \gg t_{tr}$. The gain G_{PC} is expressed as

$$G_{PC} = \frac{\tau_p}{t_{tr}} \left(1 + \frac{\mu_p}{\mu_n} \right). \tag{2.35}$$

When a photo-generated electron–hole pair is separated by an externally applied electric field, the electron crosses the detector several times before it recombines with the hole. It is noted that a larger gain results in a slower response speed, that is, the gain-bandwidth is constant in a PCD, because the gain G_{PC} is proportional to the carrier lifetime τ_p, which determines the response speed of the detector. Finally, a PCD has a relatively large dark current in general; as a PCD is essentially a conductive device, some dark current will flow. This may be disadvantage for an image sensor. Some PC materials are used as a detector overlaid on CMOS readout circuitry in a pixel due to the photoresponse with a variety of wavelengths such as X-ray, UV, and IR. Avalanche phenomena occur in some PC materials, realized in super HARP, an imaging tube with ultra high sensitivity developed in NHK[*] [102].

Several types of CMOS readout circuitry (ROC) for this purpose have been reported, see Ref. [103] for example. Another application of PCD is as replacements for on-chip color filters, described in Sec. 3.7.3 [104–106]. Some PCDs are also used for fast photodetectors; metal–semiconductor–metal (MSM) photodetectors are used for this purpose.

Metal–semiconductor–metal photodetector   The MSM photodetector is a kind of PCD, where a pair of metal fingers are placed on the surface of a semiconductor, as shown in Fig. 2.13 [86]. Because the MSM structure is easy to fabricate, MSM photodetectors are also applied to other materials such as GaAs and GaN. GaAs

---

[*]Nippon Hoso Kyokai.

MSM photodetectors are mainly used for ultra-fast photodetectors [107], although in Refs. [108, 109] GaAs MSM photodector arrays are used for image sensors. GaN MSM photodetectors have been developed for image sensors with a sensitivity the in UV region [110].

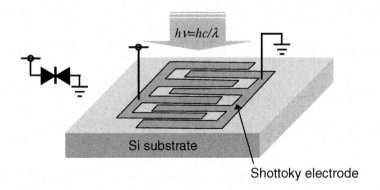

FIGURE 2.13
Structure of an MSM photodetector. The inset shows the symbols for the MSM photodetector.

2.4 Accumulation mode in PDs

A PD in a CMOS image sensor is usually operated in accumulation mode. In this mode, the PD is electrically floated and when light illuminates the PD, photocarriers are generated and swept to the surface due to the potential well in the depletion region. The PD accumulation mode was proposed and demonstrated by G.P. Weckler [13]. The potential voltage decreases when electrons accumulate. By measuring the voltage drop, the total amount of light power can be obtained. It should be noted that the accumulation of electrons is interpreted as the process of discharge in the charged capacitor by generated photocurrent.

Let us consider, using a simple but typical case, why the accumulation mode is required in a CMOS image sensor. We assume the following parameters: the sensitivity of the PD R_{ph} = 0.3 A/W, the area size of the PD A = 1000 lux, and the illumination at the PD surface L_o = 100 μm^2. Assuming that 1 lux roughly corresponds to 1.6×10^{-7} W/cm^{-2}, as described in the Appendix, the photocurrent I_{ph} is evaluated as

$$\begin{aligned} I_{ph} &= R_{ph} \times L_o \times A \\ &= 0.3 \, \text{A/W} \times 100 \times 1.6 \times 10^{-7} \, \text{W/cm}^{-2} \times 100 \, \mu\text{m}^2 \\ &\approx 10 \, \text{pA}. \end{aligned}$$

While it is possible to measure such a low photocurrent, it is difficult to precisely measure photocurrents of the same order from a two-dimensional array for a large number of points at a video rate.

2.4.1 Potential change in accumulation mode

The junction capacitance of a pn-junction PD C_{PD} is expressed as

$$C_{PD}(V) = \frac{\varepsilon_o \varepsilon_{Si}}{W}, \tag{2.36}$$

which is dependent on the applied voltage V through the dependence of the depletion width W on V, as

$$W = K(V + V_{bi})^{m_j}, \tag{2.37}$$

where K is a constant, V_{bi} is the built-in potential of the pn junction, and m_j is a parameter dependent on the junction shape: $m_j = 1/2$ for a step junction and $m_j = 1/3$ for a linear junction.

The following will hold for $C_{PD}(V)$:

$$C_{PD}(V) \frac{dV}{dt} + I_{ph} + I_d = 0, \tag{2.38}$$

where I_d is the dark current of the PD. Using Eqs. 2.36 and 2.37, Eq. 2.38 gives

$$V(t) = (V_0 + V_{bi}) \left[1 - \frac{(I_{ph} + I_d)(1 - m_j)}{C_0(V_0 + V_{bi})} t \right]^{\frac{1}{1-m_j}} - V_{bi}, \tag{2.39}$$

where V_0 and C_0 are the initial values of the voltage and capacitance in the PD, respectively. This result shows that the voltage of the PD decreases almost linearly. Usually, the PD voltage is approximately described as decreasing linearly. Figure 2.14 shows the voltage drop of a PD as a function of time. Figure 2.14 confirms that V_{PD} almost lineally decreases as time increases. Thus, light intensity can be estimated by measuring the voltage drop of the PD at a fixed time, usually at a video rate of 1/30 sec.

2.4.2 Potential description

The potential description is frequently used for CMOS image sensors and hence it is an important concept. Figure 2.15 illustrates the concept [111]. In the figure,

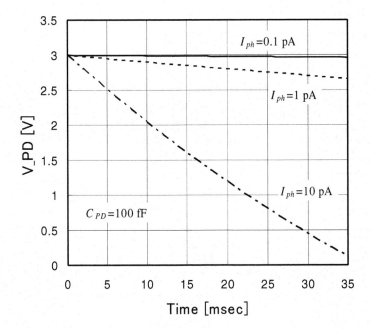

FIGURE 2.14
Voltage drop of a PD as a function of time.

a MOSFET is depicted as an example; the source acts as a PD and the drain is connected to V_{dd}. The impurity density in the source is smaller than that in the drain. The gate is off-state, or in the subthreshold region.

Figure 2.15(b) shows the potential profile along the horizontal distance, showing the conduction band edge near the surface or surface potential. In addition, the electron density at each area shown in Fig. 2.15 (c) is superimposed on the potential profile of (b); hence it is easy to see the carrier density profile, as shown in (d). The baseline of the carrier density profile sits at the bottom of the potential profile, so that the carrier density increases in the downward direction. It is noted that the potential profile or Fermi level can be determined by the carrier density; when carriers are generated by input light and accumulate in the depletion region, the potential depth changes through the change in the carrier density. However, under ordinary conditions for image sensors, the surface potential increases in proportion to the accumulated charge.

Figure 2.16 shows the potential description of a PD, which is floated electrically. This is the same situation as in the previous section, Sec. 2.4.1. In the figure, the photo-generated carriers accumulate in the depletion region of the PD. The potential well V_b is produced by the built-in potential V_{bi} plus the bias voltage V_j. Figure 2.16 (b) illustrates the accumulated state when the photo-generated carriers collect in the potential well. The accumulated charges change the potential depth from V_b to V_q,

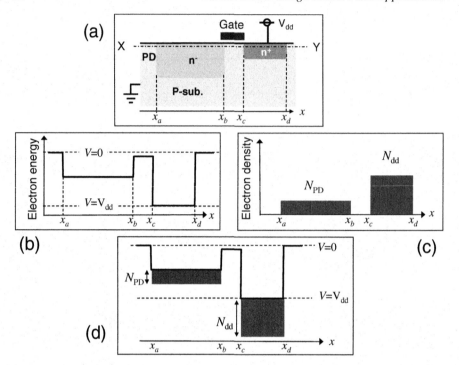

FIGURE 2.15

Illustration of potential description. An n-MOSFET structure is shown in (a), where the source is a PD and the drain is biased at V_{dd}. The gate of the MOSFET is off-state. The conduction band edge profile along X–Y in (a) is shown in (b). The horizontal axis shows the position corresponding to the position in (a) and the vertical axis shows the electron energy. The $V = 0$ and $V = V_{dd}$ levels are shown in the figure. The electron density is shown in (c) and (d) is the potential description, which is the superimposing of (a) and (c). Drawn after [111].

as shown in Figure 2.16 (b). The amount of the change $V_b - V_q$ is approximately proportional to the product of the input light intensity and the accumulation time, as mentioned in the previous section, Sec. 2.4.1.

2.4.3   Behavior of photo-generated carriers in PD

As explained in Sec. 2.3.1.2, incident photons penetrate into the semiconductor according to their energy or wavelength; photons with smaller energy or longer wavelength penetrate deep into the semiconductor, while photons with larger energy or shorter wavelength are absorbed near the surface. The photons absorbed in the depletion region are swept immediately by the electric field and accumulate in the potential well, as shown in Fig. 2.17. In Fig. 2.17, light of three colors, red, green, and blue, are incident on the PD. As shown in Fig. 2.17(a), the three lights reach differ-

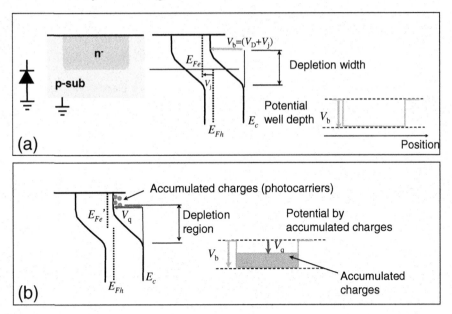

FIGURE 2.16
Potential description of a PD before accumulation (a) and after accumulation (b).

ent depths; the red light penetrates most deeply and reaches the p-substrate region, where it produces minority carrier electrons. In the p-type substrate region, there is little electric field, so that the photo-generated carriers only move by diffusion, as shown in Fig. 2.17(b). While some of the photo-generated carriers are recombined in this region and do not contribute to the signal charge, others arrive at the edge of the depletion region and accumulate in the potential well, contributing to the signal charge. The extent of the contribution depends on the diffusion length of the carriers produced in the p-substrate, electrons in this case. The diffusion length has been discussed in Sec. 2.2.2. It is noted that the diffusion length in the low impurity concentration region is large and thus carriers can travel a long distance. Consequently, the blue, green, and some portion of the red light contribute to the signal charge in this case.

This case, however, ignores the surface/interface states, which act as killers for carriers. Such states produce deep levels in the middle of the bandgap; carriers around the states are easily trapped in the levels. The lifetime in the states is generally long and trapped carriers are finally recombined there. Such trapped carriers do not contribute to the signal charge. Blue light suffers from this effect and thus has a smaller quantum efficiency than longer wavelengths.

Pinned photodiode   To alleviate the degradation of the quantum efficiency for shorter wavelengths, the pinned photodiode (PPD) or the buried photodiode (BPD) has been developed. Historically, the PPD was first developed for CCDs [112, 113],

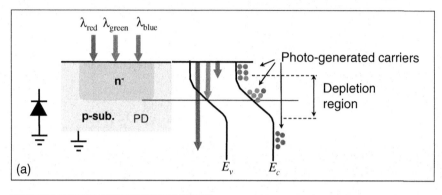

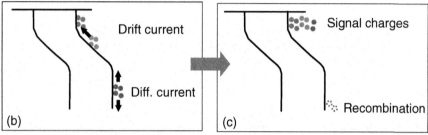

FIGURE 2.17
Behavior of photo-generated carriers in a PD.

and from the late 1990s it was adopted to CMOS image sensors [40, 114–116]. The structure of the PPD is shown in Fig. 2.19. The topmost surface of the PD has a thin p^+ layer, and thus the PD itself appears to be buried under the surface. This topmost p^+ thin layer acts to fix the Fermi level near the surface, which is the origin of the name "pinned photodiode." The p^+ layer has the same potential as the p-substrate and thus the potential profile at the surface is strongly bent so that the accumulation region is separated from the surface where the trapped states are located. In this case, the Fermi level is pinned or the potential near surface is pinned.

   Eventually, the photo-generated carriers at shorter wavelengths are quickly swept to the accumulation region by the bent potential profile near the surface and contribute to the signal charge. The PPD structure has two further merits. First, the PPD has less dark current than a conventional PD, because the surface p^+ layer masks the traps which are one of the main sources of dark current. Second, the large bent potential profile produces an accumulation region with complete depletion, which is important for 4-Tr type active pixel sensors discussed in Sec. 2.5.3. To achieve complete depletion requires not only the surface thin p^+ layer but also an elaborate design of the potential profile by precise fabrication process control. Recently, PPDs have been used for CMOS image sensors with high sensitivity.

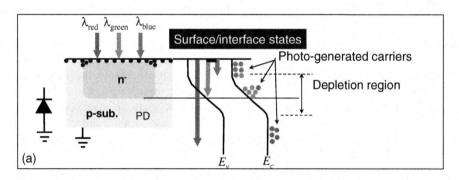

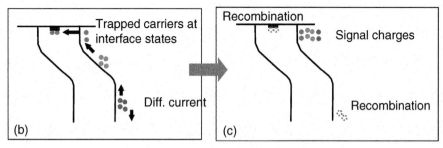

FIGURE 2.18
Behavior of photo-generated carriers in a PD with surface traps.

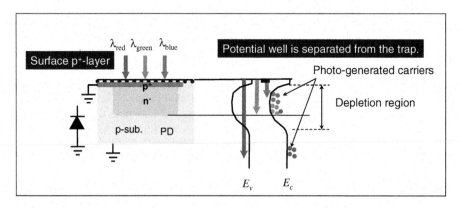

FIGURE 2.19
Behavior of photo-generated carriers in a surface p^+-layer PD or a pinned photodiode (PPD).

2.5   Basic pixel structures

In this section, basic pixel structures are described in detail. Historically, passive pixel sensors (PPS) were developed first, then active pixel sensors (APS) were developed to improve image quality. An APS has three transistors in a pixel, while a PPS has only one transistor. To achieve further improvement, an advanced APS that has four transistors in a pixel, the so-called 4T-APS, has been developed. The 4-Tr APS has greatly improved image quality, but has a very complex fabrication process. The usefulness of the 4-Tr APS is currently being debated.

2.5.1   Passive pixel sensor

PPS is a name coined to distinguish such sensors from APS, which is described in the next section. The first commercially available MOS sensor was a PPS [22, 24], but due to SNR issues, its development was halted. The structure of a PPS is very simple: a pixel is composed of a PD and a switching transistor, as shown in Fig. 2.20(a). It is similar to dynamic random access memory (DRAM).

Because of its simple structure, a PPS has a large fill factor (FF), the ratio of the PD area to the pixel area. A large FF is preferable for an image sensor. However, the output signal degrades easily. Switching noise is a crucial issue. In the first stage of PPS development, the accumulated signal charge was read as the current through the horizontal output line and then converted to a voltage through a resistance [22, 24] or a transimpedance amplifier [25]. This scheme has the following disadvantages:

- Large smear
  Smear is a ghost signal appearing as vertical stripes without any signal. A CCD can reduce smear. In a PPS, smear can occur when the signal charges are transferred into the column signal line. The long horizontal period (1H period, usually 64 μs) causes this smear.

- Large $k_B T C$ noise
  $k_B T C$ noise is thermal noise (discussed in detail in Sec. 2.7.1.2); specifically, the noise power of a charge is expressed as $k_B T C$, where C is the sampling capacitance. A PPS has a large sampling capacitance of C_C in the column signal line and hence large noise is inevitable.

- Large column FPN
  As the capacitance of the column output line C_C is large, a column switching transistor is required for a large driving capacity, and thus the gate size is large. This causes a large overlap gate capacitance C_{gd}, as shown in Fig. 2.20(a), which produces large switching noise, producing column FPN.

To address these problems, the transversal signal line (TSL) method was developed [117]. Figure 2.21 shows the concept of TSL. In the TSL structure, a column select

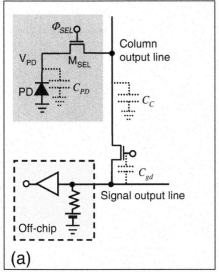

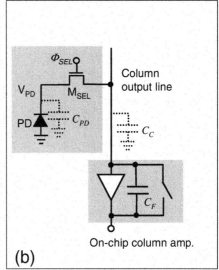

FIGURE 2.20

Basic pixel circuits of a PPS with two readout schemes. C_{PD} is a pn-junction capacitance in the PD and C_H is a stray capacitor associated with the vertical output line. In circuit (a), an off-chip amplifier is used to convert the charge signal to a voltage signal, while in circuit (b), on-chip charge amplifiers are integrated in the column so that the signal charge can be read out almost completely.

transistor is employed in each pixel. As shown in Fig. 2.21(b), signal charges are selected in every vertical period, which are much shorter than the horizontal period. This drastically reduces smear. In addition, a column select transistor M_{CSEL} required a small sampling capacitor C_{PD}, rather then the large capacitor C_C required for a standard PPS. Thus $k_B TC$ noise is reduced. Finally, the gate size M_{CSEL} can be reduced so that little switching noise occurs in this configuration. The TSL structure has also been applied to the 3T-APS in order to reduce column FPN [118].

In addition, a charge amplifier on a chip with a MOS imager in place of a resistor has been reported [119]. This configuration is effective only for a small number of pixels.

Currently, a charge amplifier placed in each column is used to completely extract the signal charge and convert it into a voltage, as shown in Fig. 2.20 (b). Although this configuration increases the performance, it is difficult to sense small signal charges due to the large stray capacitance of the horizontal output line or column output line C_C. The voltage at the column output line V_{out} is given by

$$V_{out} = Q_{PD} \frac{C_C}{C_{PD} + C_C} \frac{1}{C_F}, \tag{2.40}$$

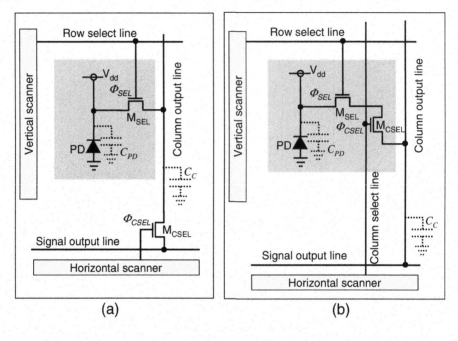

FIGURE 2.21
Improvement of PPS. (a) Conventional PPS. (b) TSL-PPS [117].

where Q_{PD} is the signal charge accumulated at the PD and C_{PD} is the capacitance of the PD.

The charge amplifier is required to precisely convert a small charge. Present C-MOS technology can integrate such charge amplifiers in each column, and thus the SNR can be improved [120]. It is noted that this configuration consumes a large amount of power.

2.5.2   Active pixel sensor, 3T-APS

The APS is named after its active element which amplifies the signal in each pixel, as shown in Fig. 2.22. This pixel configuration is called 3T-APS, compared with 4T-APS, which is described in the next section. An additional transistor M_{SF}, acts as a source follower, and thus the output voltage follows the PD voltage. The signal is transferred to a horizontal output line through a select transistor M_{SEL}. Introducing amplification at a pixel, the APS improves image quality compared with the PPS. While a PPS directly transfers the accumulated signal charges to the outside of a pixel, an APS converts the accumulated signal charges to a potential in the gate. In this configuration, the voltage gain is less than one, while the charge gain is determined by the ratio of the accumulation node charge C_{PD} to a sample and the hold node charge C_{SH}.

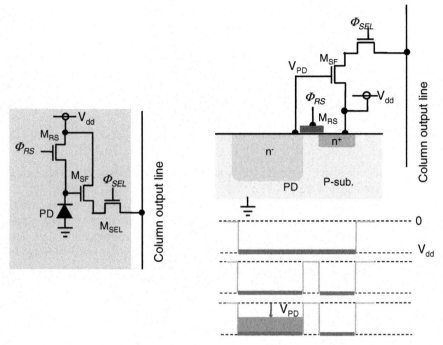

FIGURE 2.22
Basic pixel circuits of a 3T-APS.

The operation of an APS is as follows. First, the reset transistor M_{RS} is turned on. Then the PD is reset to the value $V_{dd} - V_{th}$, where V_{th} is the threshold voltage of transistor M_{RS} (see Fig. 2.22(c)). Next, M_{RS} is turned off and the PD is electrically floated (Fig. 2.22(d)). When light is incident, the photo-generated carriers accumulate in the PD junction capacitance C_{PD} (Fig. 2.22). The accumulated charge changes the potential in the PD; the voltage of the PD V_{PD} decreases according to the input light intensity, as described in Sec. 2.4.1. After an accumulation time, for example, 33 msec at video rate, the select transistor M_{SEL} is turned on and the output signal in the pixel is read out in the vertical output line. When the read-out process is finished, M_{SEL} is turned off and M_{RS} is again turned on to repeat the above process.

It is noted that the accumulated signal charge is not destroyed, which make it possible to read the signal multiple times. This is a useful characteristic for smart CMOS image sensors.

2.5.2.1 Issues with 3T-APS

Although the APS overcomes the disadvantage of the PPS, namely low SNR, there are several issues with the APS, as follows:

- It is difficult to suppress $k_B TC$ noise.

- The photodetection region, that is, the PD, simultaneously acts as a photoconversion region. This constrains the PD design.

Here we define the terms of full-well capacity and conversion gain. The full-well capacity is the number of charges that can be accumulated in the PD. The larger the full-well capacity, the wider the dynamic range (DR), which is defined as the ratio of the maximum output signal value V_{max} to the detectable signal value V_{min}:

$$DR = 20\log\frac{V_{max}}{V_{min}} \text{ [dB]}. \qquad (2.41)$$

The conversion gain is defined as the voltage change when one charge (electron or hole) is accumulated in the PD. The conversion gain is thus equal to $1/C_{PD}$.

The full-well capacity increases as the PD junction capacitance C_{PD} increases, while the conversion gain, which is a measure of the increase of the PD voltage according to the amount of accumulated charge, is inversely proportional to C_{PD}. This implies that the full-well capacity and the conversion gain have a trade-off relationship in a 3T-APS. The 4T-APS resolves the trade-off as well as suppressing k_BTC noise.

2.5.3 Active pixel sensor, 4T-APS

To alleviate the issues with the 3T-APS, the 4T-APS has been developed. In a 4T-APS, the photodetection and photoconversion regions are separated. Thus, the accumulated photo-generated carriers are transferred to a floating diffusion (FD) where the carriers are converted to a voltage. One transistor is added to transfer charge accumulated in the PD to the FD, making the total number of transistors in a pixel four, and hence this pixel configuration is called 4T-APS. Figure 2.23 shows the pixel structure of the 4T-APS.

The operation procedure is as follows. First, the signal charge accumulates in the PD. It is assumed that in the initial stage, there is no accumulated charge in the PD; a condition of complete depletion is satisfied. Just before transferring the accumulated signal charge, the FD is reset by turning on the reset transistor M_{RS}. The reset value is read out for correlated double sampling (CDS) to turn on the select transistor M_{SEL}. After the reset readout is finished, the signal charge accumulated in the PD is transferred to the FD by turning on the FD with a transfer gate M_{TG}, following the readout of the signal by turning on M_{SEL}. Repeating this process, the signal charge and reset charge are read out. It is noted that the reset charge can be read out just after the signal charge readout. This timing is essential for CDS operation and can be realized by separating the charge accumulation region (PD) and the charge readout region (FD); this timing eliminates k_BTC noise and it cannot be achieved by the 3T-APS. By this CDS operation, the 4T-APS achieves low noise operation and thus is performance is comparable to CCDs. It is noted that in the 4T-APS the PD must be drained of charge completely in the readout process. For this, a PPD is required. A carefully designed potential profile can achieve a complete transfer of accumulated charge to the FD through the transfer gate.

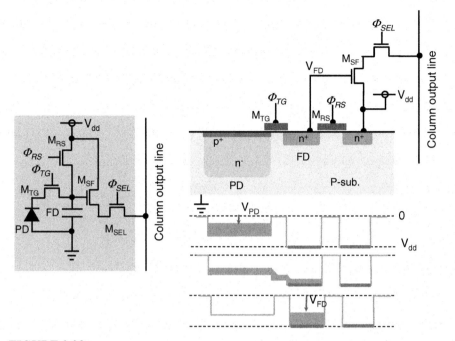

FIGURE 2.23

Basic pixel circuits of the 4T-APS.

2.5.3.1 Issues with 4T-APS

Although the 4T-APS is superior to the 3T-APS in its low noise level, there are some issues with the 4T-APS, as follows:

- The additional transistor reduces the FF compared with the 3T-APS.

- Image lag may occur when the accumulated signal charge is completely transferred into the FD.

- It is difficult to establish fabrication process parameters for the PPD, transfer gate, FD, reset transistor, and other units, for low noise and low image lag performance.

Figure 2.24 illustrates incomplete charge transfer in a 4T-APS. In Fig. 2.24 (a), the charges are completely transferred to the FD, while in Fig. 2.24 (b) some charge remains in the PD, causing image lag. To prevent incomplete transfer, elaborate potential profiles are required [121, 122].

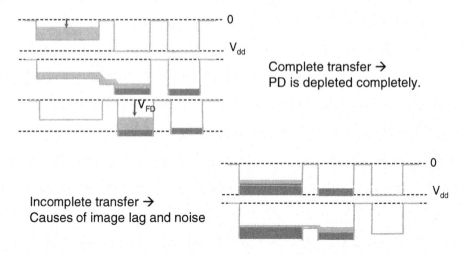

Complete transfer →
PD is depleted completely.

Incomplete transfer →
Causes of image lag and noise

FIGURE 2.24
Incomplete charge transfer in a 4T-APS.

2.6   Sensor peripherals

2.6.1   Addressing

In CMOS image sensors, to address each pixel, a scanner or a decoder is used. A scanner consists of a latch array or shift register array to carry data in accordance with a clock signal. When using scanners with vertical and horizontal access, the pixels are sequentially addressed. To access an arbitrary pixel, a decoder, which is a combination of logic gates, is required.

A decoder arbitrarily converts N input data to 2^N output data using customized random logic circuits. Figure 2.25 shows a typical scanner and a decoder. Figure 2.26 presents an example of a decoder, which decodes 3-bit input data to 6 output data.

Random access   An advantage of smart CMOS image sensors is random access capability, where an arbitrary pixel can be addressed at any time. The typical method to implement random access is to add one transistor to each pixel so that a pixel can be controlled with a column switch, as shown in Fig. 2.27. Row and column address decoders are also required instead of scanners, as mentioned above. It is noted that if the extra transistor is added in series with the reset transistor, as shown in Fig. 2.28, then anomalies will occur for some timings [123]. In this case, if M_{RRS} is turned on, the accumulated charge in the PD is distributed between the PD capacitance C_{PD} and a parasitic capacitance C_{diff}, which degrades the signal charge.

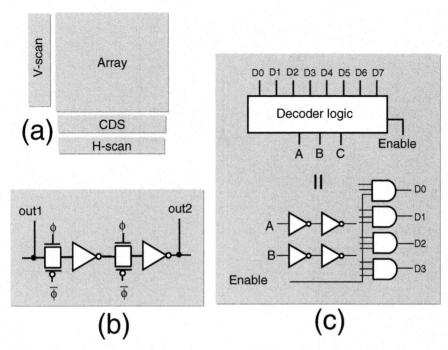

FIGURE 2.25
Addressing methods for CMOS image sensors: (a) sensor architecture, (b) scanner, (c) decoder.

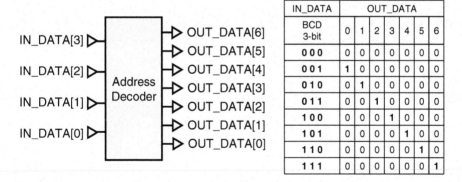

FIGURE 2.26
Example of a decoder.

Multiresolution Multiresolution is another addressing technique for CMOS image sensors [68, 124]. Mulitresolution is a method to vary the resolution in a sensor; for example, in a VGA (640 × 480-pixel) sensor, the resolution can be changed by a factor of 1/4 (320 × 240-pixel), a factor of 1/8 (160 × 120-pixel), and so on. To

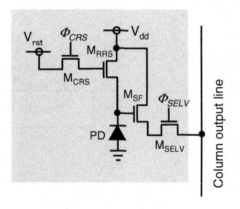

FIGURE 2.27
Pixel structure for random access.

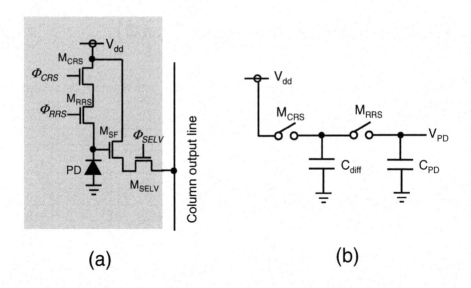

FIGURE 2.28
(a) Pixel structure for random access for the different types shown in Fig. 2.27. (b) Equivalent circuits.

quickly locate an object with a sensor, a coarse resolution is effective as the post-image processing load is low. This is effective for target tracking, robotics, etc.

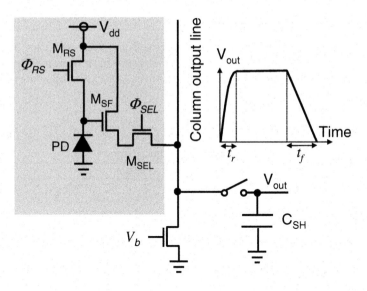

FIGURE 2.29

Readout circuits using a source follower. The inset shows the output voltage V_{out} dependence on the time in the readout cycle [125].

2.6.2 Readout circuits

2.6.2.1 Source follower

The voltage of a PD is read with a source follower (SF). As shown in Fig. 2.29, a follower transistor M_{SF} is placed in a pixel and a current load M_b is placed in each column. A select transistor M_{SEL} is located between the follower and the load. It is noted that the voltage gain A_v of an SF is less than 1 and is expressed by the following:

$$A_v = \frac{1}{1 + g_{mb}/g_m},\qquad(2.42)$$

where g_m and g_{mb} are the transconductance and the body transconductance of M_{SF}, respectively [126]. The DC response of an SF is not linear over the input range. The output voltage is sampled and held in the capacitance C_{CDS}.

In the readout cycle using an SF, the charge and discharge processes associated with an S/H capacitor C_{SH} are the same. In the charge process, C_{SH} is charged with a constant voltage mode so that the rise time t_r is determined by the constant voltage mode. In the discharge process, CH_H is discharged with a constant current mode by the current source of the SF so that the fall time t_f is determined by the constant current mode. The inset of Fig. 2.29 illustrates this situation. These characteristics must be evaluated when the readout speed is important [125].

2.6.2.2 Correlated double sampling

CDS is used to eliminate thermal noise generated in a reset transistor of the PD, which is $k_B TC$ noise. Several types of CDS circuitry have been reported and are reviewed in Ref. [3] in detail. Table 2.2 summarizes CDS types following the classification of Ref. [3].

TABLE 2.2

CDS types for CMOS image sensors

Category	Method	Feature	Ref.
Column CDS 1	One coupling capacitor	Simple structure but suffers from column FPN	[127]
Column CDS 2	Two S/H capacitors	ibid.	[128]
DDS*	DDS following column CDS	Suppression of column FPN	[129]
Chip-level CDS	I–V conv. and CDS in a chip	Suppression of column FPN but needs fast operation	[116]
Column ADC	Single slope ADC	Suppression of column FPN	[130, 131]
	Cyclic ADC	ibid.	[132]

*double delta sampling.

Figure 2.30 shows typical circuitry for CDS with an accompanying 4T-APS type pixel circuit. The basic CDS circuit consists of two sets of S/H circuits and a differential amplifier. The reset and signal level are sampled and held in the capacitances C_R and C_S, respectively, and then the output signal is produced by differentiating the reset and signal values held in the two capacitors. The operation principle can be explained as follows with the help of the timing chart in Fig. 2.31 and with Fig. 2.30. In the signal readout phase, the select transistor M_{SEL} is turned on from t_1 to t_7 when Φ_{SEL} turns on ("HI"–level). The first step is to read the reset level or $kT_B C$ noise and store it in capacitor C_R just after the FD is reset at t_2 by setting Φ_{RS} to HI. To sample and hold the reset signal in the capacitor C_R, Φ_R becomes HI at t_3. The next step is to read the signal level. After transferring the accumulated signal charge to the FD by turning on the transfer gate of M_{TG} at t_4, the accumulated signal is sampled and held in C_S by setting Φ_S to HI. Finally, the accumulated signal and the reset signal are differentiated by setting Φ_Y to HI.

Another CDS circuit is shown in Fig. 2.32 [127, 133]. In this case, the capacitor C_1 is used to subtract the reset signal.

2.6.3 Analog-to-digital converters

In this section, analog-to-digital converters (ADCs) for CMOS image sensors are briefly described. For sensors with a small number of pixels, such as QVGA (230×320) and CIF (352×288), a chip-level ADC is used [134], [135]. When the number of

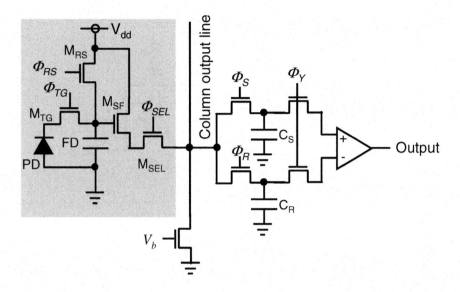

FIGURE 2.30
Basic circuits of CDS.

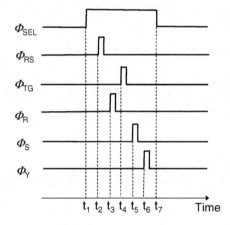

FIGURE 2.31
Timing chart of CDS. The symbols are the same as in Fig. 2.30.

pixels increases, column parallel ADCs are employed, such as a successive approximation ADC [136, 137], a single slope ADC [66, 130, 131, 138], and a cyclic AD-C [132, 139]. Also, pixel-level ADCs have been reported [61, 140, 141]. The place where an ADC is employed is the same point of view in the architecture of smart CMOS image sensors, that is, pixel-level, column-level, and chip-level.

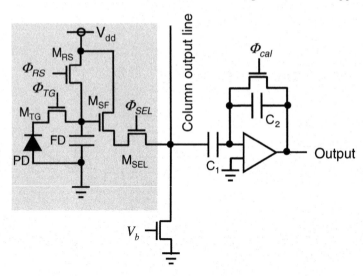

FIGURE 2.32
An alternative circuit for CDS. Here a capacitor is used to subtract the reset signal.

2.7   Basic sensor characteristics

In this section, some basic sensor characteristics are described. For details on measurement techniques of image sensors, please refer to Refs. [142, 143].

2.7.1   Noise

2.7.1.1   Fixed pattern noise

In an image sensor, spatially fixed variations of the output signal are of great concern for image quality. This type of noise is called fixed pattern noise (FPN). Regular variations, such as column FPN, can be perceived more easily than random variations. A variation of 0.5% of pixel FPN is an acceptable threshold value, while 0.1% of column FPN is acceptable [144]. Employing column amplifiers sometimes causes column FPN. In Ref. [144], column FPN is suppressed by randomizing the relation between the column output line and the column amplifier.

2.7.1.2 $k_B TC$ noise

In a CMOS image sensor, the reset operation mainly causes thermal noise. When the accumulated charge is reset through a reset transistor, the thermal noise $4k_B T R_{on} \delta f$ is sampled in the accumulation node, where δf is the frequency bandwidth and R_{on} is the ON-resistance of the reset transistor, as shown in Fig. 2.33. The accumulation node is a PD junction capacitance in a 3T-APS and an FD capacitance in a 4T-APS.

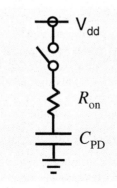

FIGURE 2.33
Equivalent circuits of kTC noise. R_{on} is the ON-resistance of the reset transistor and C_{PD} is the accumulation capacitance, which is a PD junction capacitance for a 3T-APS and a floating diffusion capacitance for a 4T-APS, respectively.

The thermal noise is calculated to be $k_B T/C_{PD}$, which does not depend on the ON-resistance R_{on} of the reset transistor. This is because larger values of R_{on} increase the thermal noise voltage per unit bandwidth while they decrease the bandwidth [145], which masks the dependence of R_{on} on the thermal noise voltage. We now derive this formula referring to the configuration in Fig. 2.33. The thermal noise voltage is expressed as

$$\overline{v_n^2} = 4k_B T R_{on} \Delta f. \tag{2.43}$$

As shown in Fig. 2.33, the transfer function is expressed as

$$\frac{v_{out}}{v_n}(s) = \frac{1}{R_{on}C_{PD}s + 1}, \quad s = j\omega. \tag{2.44}$$

Thus, the noise is calculated as

$$\begin{aligned}
\overline{v_{out}^2} &= \int_0^\infty \frac{4k_B T R_{on}}{(2\pi R_{on} Cf)^2 + 1} df \\
&= \frac{k_B T}{C}.
\end{aligned} \tag{2.45}$$

The noise power of the charge q_{out}^2 is expressed as

$$q_{out}^2 = (Cv_{out})^2 = k_B TC. \tag{2.46}$$

The term "kTC" noise originates from this formula. The $k_B TC$ noise can be eliminated by the CDS technique, though it can only be applied to a 4T-APS, as it is difficult to apply to a 3T-APS.

2.7.1.3   Reset method

The usual reset operation in a 3T-APS is to turn on M_{rst} (shown in Fig. 2.34 (a)) by applying a voltage of HI or V_{dd} to the gate of M_{rst} and to fix the voltage of the PD V_{PD} at $V_{dd} - V_{th}$, where V_{th} is the threshold voltage of M_{rst}. It is noted that in the final stage of the reset operation, V_{PD} reaches $V_{dd} - V_{th}$, so that the gate–source voltage across M_{rst} becomes less than V_{th}. This means that M_{rst} enters the subthreshold region. In this sate, V_{PD} slowly reaches $V_{dd} - V_{th}$. This reset action is called a soft reset [146]. By employing PMOSFET with the reset transistor, this Problem can be avoided, although PMOSFET consumes more area than NMOSFET because it needs an Nwell area. In contrast, in a hard reset, the applied gate voltage is larger than V_{dd}, and thus M_{rst} is always above the threshold, so that the reset action finishes quickly. In this case, $k_B TC$ noise occurs as previously mentioned.

A soft reset has the disadvantage of causing image lag, while it has the advantage of reducing $k_B TC$ noise; the noise voltage is equal to $\sqrt{k_B T/2C}$ [146]. By combining a soft reset and a hard reset, $k_B TC$ noise can be reduced and image lag suppressed, which is called a flushed reset [147], as shown in Fig. 2.34. In a flushed reset, the PD is first reset by a hard reset to flush the accumulated carriers completely. It is then reset by a soft reset to reduce $k_B TC$ noise. A flush reset requires a switching circuit to alternate the bias voltage of the gate in the reset transistor.

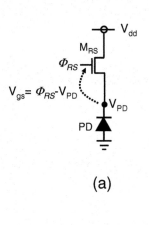

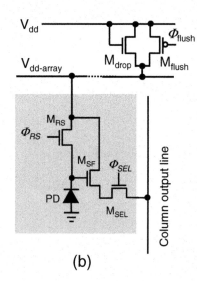

FIGURE 2.34
(a) Reset operation in a 3T-APS and (b) a flushed reset [147].

2.7.2 Dynamic range

The dynamic range (DR) of an image sensor is defined as the ratio of the output signal range to the input signal range. DR is thus determined by two factors, the noise floor and the well charge capacity. The most of sensors have almost the same DR of around 70 dB, which is mainly determined by the well capacity of the PD. For some applications, such as in automobiles, this value of 70 dB is not sufficient; a DR of over 100 dB is required for these applications. Considerable effort to enhance DR has been made and is described in Chapter 5.

2.7.3 Speed

The speed of an APS is basically limited by the diffusion carriers. Some of the photo-generated carriers in the deep region of a substrate will finally arrive at the depletion region, acting as slow output signals. The diffusion time for electrons and holes as a function of impurity concentration is shown in Fig. 2.5. Is it noted that the diffusion lengths for both holes and electrons are over a few tens of micrometers and sometimes reach a hundred micrometers, and careful treatment is needed to achieve high speed imaging. This effect greatly degrades the PD response, especially in the IR region. To alleviate this effect, some structures prevent diffusion carriers from entering the PD region.

CR time constants are another major factor limiting the speed, because the vertical output line is generally so long in smart CMOS image sensors that the associated resistance and stray capacitance are large.

It is noted that the total of the overlap capacitances of the select transistors in the pixels connected to the vertical output line is large and thus it cannot be ignored compared with the stray capacitors in the vertical output line.

2.8 Color

There are three ways to realize color in a conventional CMOS image sensor, as shown in Fig. 2.35.

They are explained as follows:

On-chip color filter type Three colored filters are directly placed on the pixels, typically red (R), green (G), and blue (B) (RGB) or CMY complementary color filters of cyan (Cy), magenta (Mg), and yellow (Ye) and green are used. The representation of CMY and RGB is as follows (W indicates white):

$$
\begin{aligned}
Ye &= W - B = R + G, \\
Mg &= W - G = R + B, \\
Cy &= W - R = G + B.
\end{aligned}
\tag{2.47}
$$

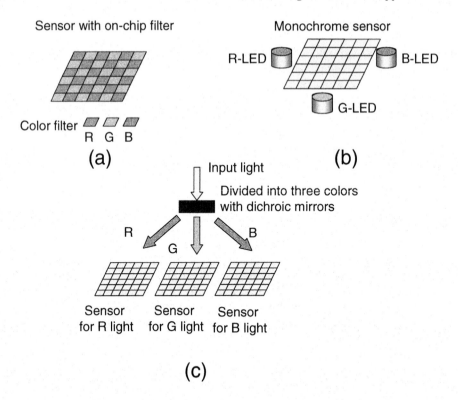

FIGURE 2.35
Methods to realize color in CMOS image sensors. (a) On-chip color filter. (b) Three image sensors. (c) Three light sources.

The Bayer pattern is commonly used to place the three RGB filters [148]. This type of on-chip filter is widely used in commercially available CMOS image sensors. Usually, the color filters are organic film, but inorganic color film has also been used [149]. The thickness of α-Si is controlled to produce a color response. This helps reduce the thickness of the color filters, which is important in optical crosstalk in a fine pitch pixel less than 2 μm in pitch.

Three imagers type   In the three imagers method, three CMOS image sensors without color filters are used for the R, G, and B colors. To divide the input light into three colors, two dichroic mirrors are used. This configuration realizes high-color fidelity but requires complicated optics and is expensive. It is usually used in broadcasting systems, which require high-quality images.

Three light sources type   The three light sources method uses artificial RGB light sources, with each RGB source illuminating the objects sequentially. One sensor acquires three images for the three colors, with the three images combined to form the final image. This method is mainly used in medical endoscopes. The color fidelity is excellent but the time to acquire a whole image is longer than for the above two methods. This type of color representation is not applicable to conventional CMOS image sensors because they usually have a rolling shutter. This is discussed in Sec. 5.4.

Although color is an important characteristic for general CMOS image sensors, the implementation method is almost the same as that for smart CMOS image sensors and a detailed discussion is beyond the scope of this book. Color treatment for general CMOS image sensors is described in detail in Ref. [2]. Section 3.7.3 details selected topics on realizing colors using smart functions.

2.9 Pixel sharing

Some parts in a pixel for example FD, can be shared each other, so that the pixel size can be reduced [150]. Figure 2.36 shows some examples of pixel sharing schemes. The FD driving sharing technique [153] shown in Fig. 2.36(d) is used to reduce the number of transistors in a 4T-APS by one [154]. The select transistor can be eliminated by controlling the potential of the FD by changing the pixel drain voltage through the reset transistor. Recently, sensors have been reported with around 2-μm pitch pixels using pixel sharing technology [155, 156]. In Ref. [156], a zigzag placement of RGB pixels improves the configuration for pixel sharing, as shown in Fig. 2.37.

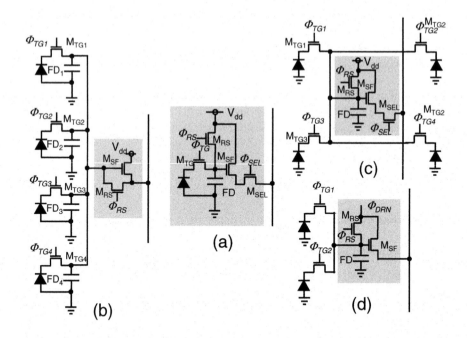

FIGURE 2.36

Pixel sharing. (a) Conventional 3T-APS. (b) Sharing of a select transistor and a source follower transistor [151]. (c) Pixels with only a PD and transfer gate transistor while the other elements including the FD are shared [152]. (d) As in (c) but with the reset voltage controlled [153].

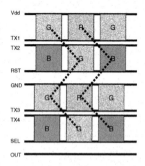

FIGURE 2.37

Pixel sharing with a zigzag placement of RGB pixels [156].

2.10 Comparison between pixel architecture

In this section, several types of pixel architecture, PPS, 3T-APS, and 4T-APS, are summarized in Table 2.3, as well as a log sensor which is detailed in Chapter 3. At present, the 4T-APS has the best performance with regard to noise characteristics and will eventually become widely used in CMOS image sensors. However, it should be noted that other systems have advantages, which provide possibilities for smart sensor functions.

TABLE 2.3
Comparison between PPS, 3T-APS, 4T-APS, and log sensor. The log sensor is discussed in Chapter 3.

	PPS	3T-APS	4T-APS (PD)	4T-APS (PG)	Log
Sensitivity	Depends on the performance of a charge amp	Good	Good	Fairly good	Good but poor at low light level
Area consumption	Excellent	Good	Fairly good	Fairly good	Poor
Noise	Fairly good	Fairly good (no kTC reduction)	Excellent	Excellent	Poor
Dark current	Good	Good	Excellent	Good	Fairly good
Image lag	Fairly good	Good	Fairly good	Fairly good	Poor
Process	Standard	Standard	Special	Special	Standard
Note	Very few commercialized	Widely commercialized	Widely commercialized	Very few commercialized	Recently commercialized

2.11 Comparison with CCDs

In this section, CMOS image sensors are compared with CCDs. The fabrication process technologies of CCD image sensors have been developed only for CCD image sensors themselves, while those of CMOS image sensors were originally developed for standard mixed signal processes. Although the recent development of CMOS image sensors requires dedicated fabrication process technologies, CMOS image sensors are still based on standard mixed singal processes.

There are two main differences between the architecture of CCD and CMOS sensors, the signal transferring method and the signal readout method. Figure 2.38 illustrates the structures of CCD and CMOS image sensors. A CCD transfers the

TABLE 2.4

Comparison between a CCD image sensor and a CMOS image sensor

Item	CCD image sensor	CMOS image sensor
Readout scheme	One on-chip SF; limits speed	SF in every column; may exhibit column FPN
Simultaneity	Simultaneous readout of every pixel	Sequential reset for every row; rolling shutter
Transistor isolation	Reverse biased pn-junction	LOCOS/STI[†]; may exhibit stress-induced dark currents
Thickness of gate oxide	Thick for complete charge transfer (> 50 nm)	Thin for high speed transistor and low voltage power supply (< 10 nm)
Gate electrode	Overlapped 1st & 2nd poly-Si layers	Polycide poly-Si
Isolation layers	Thin for suppressing light guide	Thick ($\sim 1\mu$m)
Metal layer	Usually one	Over three layers

[†]LOCOS: local oxidation of silicon, STI: shallow trench isolation.

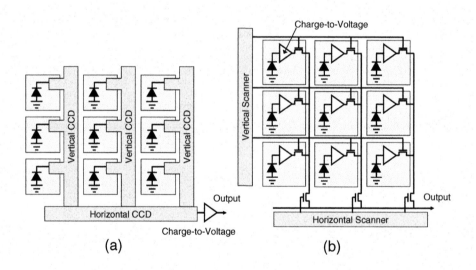

(a)   (b)

FIGURE 2.38

Conceptual illustration of the chip structure of (a) CCD and (b) CMOS image sensors.

signal charge to the end of the output signal line as it is and converts it into a voltage signal through an amplifier. In contrast, a CMOS image sensor converts the signal charge into a voltage signal at each pixel. The in-pixel amplification may cause FPN and thus the quality of early CMOS image sensors was worse than that of CCDs.

However, this problem has been drastically improved. In high-speed operation, the in-pixel amplification configuration gives better gain-bandwidth than a configuration with one amplifier on a chip.

In CCD image sensors, the signal charge is transferred simultaneously, which gives low noise and high power consumption. Also, this signal transfer gives the same accumulation time for every pixel at any time. In contrast, in CMOS image sensors, the signal charge is converted at each pixel and the resultant signal is read out row-by-row, so that the accumulation time is different for pixels in different rows at any time. This is referred to as a "rolling shutter." Figure 2.39 illustrates the origin of the rolling shutter. A triangle shape object moves from left to right. In the imaging plane, the object is scanned row by row. In Fig. 2.39(a) at $Time_k$ (k = 1, 2, 3, 4, and 5), the sampling points are shown in Row #1–#5. The original figure (left in Fig. 2.39 (b)) is distorted in the detected image (right in Fig. 2.39 (b)), which is constructed from the corresponding points in Fig. 2.39 (a). Table 2.4 summarizes the comparison between CCD and CMOS image sensors, including these features.

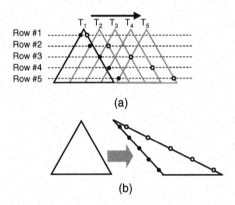

(a)

(b)

FIGURE 2.39
Illustration of the origin of a rolling shutter. (a) A triangle-shape object moves from left to right. (b) The original image is distorted.

3

Smart functions and materials

3.1 Introduction

A CMOS image sensor can employ smart functions on its chip. In this chapter, several smart functions and materials are overviewed. First, smart functions are described, then in the following sections, the output from pixels is discussed.

In a conventional CMOS image sensor, the output from a pixel is a voltage signal with a source follower (SF), giving an analog output, as mentioned in Sec. 2.6.2.1. To realize some smart functions, however, other kinds of modes such as the current mode with analog, digital, or pulse processing modes have been developed, as summarized in Table 3.1. In Sec. 3.2.1, two types of current mode operation are explained. Processing methods are then discussed. Analog processing is introduced first. Pulse mode processing is then presented in detail. Pulse mode processing is a mixture of analog and digital processing. Finally, details on digital processing are presented.

TABLE 3.1

Signal processing categories for smart CMOS image sensors

Category	Pros	Cons
Analog	Easy to realize sum and subtraction operation	Difficult to achieve high precision and programmability
Digital	High precision and programmability	Difficult to achieve a good FF*
Pulse	Medium precision, easy to process signals	Difficult to achieve a good FF

*fill factor.

In the latter part of this chapter, structures and materials other than standard silicon CMOS technologies are introduced for certain CMOS image sensors. The recent advancement of LSI technologies has introduced many new structures such as silicon-on-insulator (SOI), silicon-on-sapphire (SOS), and three-dimensional integration, and the use of many other materials such as SiGe and Ge. The use of these new structures and materials in a smart CMOS image sensor can enhance its performance and functions.

TABLE 3.2
Structures and materials for smart CMOS image sensors

Structure/material	Features
SOI	Small area when using both NMOS and PMOS
SOS	Transparent substrate (sapphire)
3D integration	Large FF
SiGe/Ge	Long wavelength (NIR)

3.2 Pixel structure

In this section, various pixel structures for smart CMOS image sensors are introduced. These structures are different from the conventional active pixel sensor (APS) structure, but they can be useful for smart functions.

3.2.1 Current mode

The conventional APS outputs a signal as a voltage. For signal processing, the current mode is more convenient, because signals can easily be summed and subtracted by Kirchoff's current law. For an arithmetic unit, multiplication can be easily employed with a current mirror circuit, which is also used to multiply the photocurrent through a mirror ratio larger than one. It is noted however that this causes pixel fixed pattern noise (FPN).

In the current mode, memory can be implemented using a current copier circuit [157]. FPN suppression and analog to digital converters (ADCs) in current mode have also been demonstrated [158]. The current mode is classified into two categories: direct output mode and accumulation mode.

3.2.1.1 Direct mode

In the direct output mode, the photocurrent is directly output from the photodetector: photodiode (PD) or phototransistor (PTr) [159, 160]. The photocurrent from a PD is usually transferred using current mirror circuits with or without current multiplication. Some early smart image sensors used current mode output with phototransistor by a current mirror. The mirror ratio is used to amplify the input photocurrent, as described above. The architecture however suffers from low sensitivity at low light level and large FPN due to mismatch in the current mirror. The basic circuit is shown in Fig. 3.1.

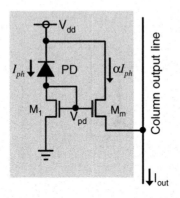

FIGURE 3.1
Basic circuit of a pixel using a current mirror. The ratio of the transistor W/L in M_1 to M_m is α, so that the mirror current or the output current is equal to αI_{ph}.

3.2.1.2 Accumulation mode

Figure 3.2 shows the basic pixel structure of a current mode APS [158, 161, 162]. By introducing the APS configuration, the image quality is improved compared with the direct mode. The output of the pixel is expressed as follows:

$$I_{pix} = g_m \left(V_{gs} - V_{th}\right)^2, \tag{3.1}$$

where V_{gs} and g_m are the gate-source voltage and the transconductance of the transistor M_{SF}, respectively. At the reset, the voltage at the PD node is

$$V_{reset} = \sqrt{\frac{2L_g}{\mu_n C_{ox} W_g} I_{ref}} + V_{th}. \tag{3.2}$$

As light is incident on the PD, the voltage at the node becomes

$$V_{PD} = V_{reset} - \Delta V, \tag{3.3}$$

where T_{int} is the accumulation time and ΔV is

$$\Delta V = \frac{I_{ph} T_{int}}{C_{PD}}, \tag{3.4}$$

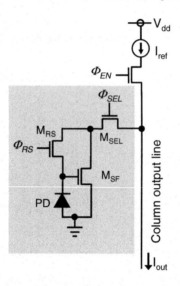

FIGURE 3.2
Basic circuit of a current mode APS [161].

which is the same as for a voltage mode APS. Consequently, the output current is expressed as

$$I_{pix} = \frac{1}{2}\mu_n C_{ox}\frac{W_g}{L_g}\left(V_{reset} - \Delta V - V_{th}\right)^2. \tag{3.5}$$

The difference current $I_{diff} = I_{ref} - I_{pix}$ is then

$$I_{diff} = \sqrt{2\mu_n C_{ox}\frac{W_g}{L_g}I_{ref}\Delta V - \frac{1}{2}\mu_n C_{ox}\frac{W_g}{L_g}\Delta V^2}. \tag{3.6}$$

It is noted that the threshold voltage of the transistor M_{SF} is cancelled so that the FPN originating from the variation of the threshold should be improved. Further discussion appears in Refs. [161, 162].

3.2.2   Log sensor

A conventional image sensor responds linearly to the input light intensity. A log sensor is based on the subthreshold operation mode of MOSFET. Appendix E explains the subthreshold operation. A log sensor pixel uses the direct current mode, because the current mirror configuration is a log sensor structure when the photocurrent is so small that the transistor enters the subthreshold region. Another application of log sensors is for wide dynamic range image sensors [163–168]. Wide dynamic range image sensors are described in Sec. 4.4.

Figure 3.3 shows the basic pixel circuit of a logarithmic CMOS image sensor. In the subthreshold region, the MOSFET drain current I_d is very small and exponentially increases with gate voltage V_g:

$$I_d = I_o \exp\left(\frac{e}{mk_B T}(V_g - V_{th})\right). \tag{3.7}$$

For the derivation of this equation and the meaning of the parameters, refer to Appendix E.

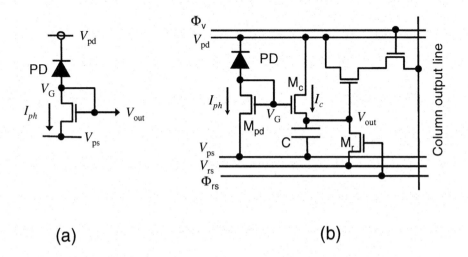

(a)   (b)

FIGURE 3.3
Pixel circuit of a log CMOS image sensor. (a) Basic pixel circuit. (b) Circuit including accumulation mode [169].

In the log sensor of Fig. 3.3(b),

$$V_G = \frac{mkT}{e}\ln\left(\frac{I_{ph}}{I_o}\right) + V_{ps} + V_{th}. \tag{3.8}$$

In this sensor the accumulation mode is employed in the log sensor architecture. For a drain current of M_c, I_c is expressed as

$$I_c = I_o \exp\left[\frac{e}{mk_B T}(V_G - V_{out} - V_{th})\right]. \tag{3.9}$$

This current I_c is charged to the capacitor C and thus the time variation of V_{out} is given by

$$C\frac{dV_{out}}{dt} = I_c dt. \tag{3.10}$$

By substituting Eq. 3.8 into Eq. 3.9, we obtain the following equation:

$$I_c = I_{ph} \exp \left[\frac{e}{mk_B T} (V_{out} - V_{ps}) \right].$$ (3.11)

By substituting this into Eq. 3.10 and integrating, the output voltage V_{out} is obtained as

$$V_{out} = \frac{mkT}{e} \ln \left(\frac{e}{mkTC} \int I_{ph} dt \right) + V_{ps}.$$ (3.12)

Although a log sensor has a wide dynamic range over 100 dB, it has some disadvantages, such as low photosensitivity especially in the low illumination region compared with a 4T-APS, slow response due to subthreshold operation, and a relatively large variation of the device characteristics due to subthreshold operation.

3.3 Analog operation

In this section, some basic analog operations are introduced.

3.3.1 Winner-take-all

Winer-take-all (WTA) circuits are a type of analog circuit [170]. Figure 3.4 shows W-TA circuits with N current inputs. Each WTA cell consists of two MOSFETs, $M_{i(k)}$ and $M_{e(k)}$, with a mutually connected gate-drain [171]. Here we only consider two WTA cells, the k-th and $(k+1)$-th cells. The key feature to understand in the WTA operation principle is that the transistor $M_{i(k)}$ operates in the saturation region with channel length modulation and acts as an inhibitory feedback, while the transistor $M_{e(k)}$ operates in the subthreshold region and acts as an excitatory feedforward.

Consider the situation where the input current $I_{in(k)}$ is a little larger than $I_{in(k+1)}$, with a difference of δI. In the initial state, the input currents of the k-th and $(k+1)$-th cells are the same. Then the input current of the k-th cell gradually increases. Because of the channel modulation effect (see Appendix E), $V_{d(k)}$ increases as $I_{in(k)}$ increases. As $V_{d(k)}$ is also the gate voltage of the transistor $M_{e(k)}$, which operates in the subthreshold region, the drain current $I_{out(k)}$ increases exponentially. As $\sum_{i=1}^{N} I_{out(i)}$ is constant and equal to I_b, this exponential increase of $I_{out(k)}$ causes the other currents $I_{out(i)}$, $i \neq k$ to diminish. Eventually, only the input current $I_{in(k)}$ flows, that is, the winner-take-all action is achieved.

WTA is an important technique for analog current operation, because it automatically calculates the maximum of a number of inputs in parallel. The disadvantage of WTA is its relatively slow response due to subthreshold operation.

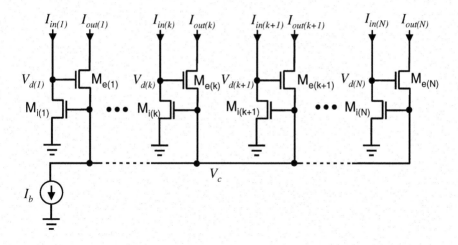

FIGURE 3.4
WTA circuits [171].

3.3.2 Projection

Projection is a method to project pixel values in one direction, usually the row and column directions as shown in Fig. 3.5. It results in the compression of data from $M \times N$ to $M + N$, where M and N are column and row number, respectively. It is a useful preprocess method for image processing, because it is simple and fast [172]. In current mode, the projection operation is easily achieved by summing the output current in the horizontal or vertical directions [173, 174].

3.3.3 Resistive network

The silicon resistive network, first proposed by C. Mead [42], is inspired by biological signal processing systems, where massive parallel processing is achieved in real time with ultra low power consumption. A detailed analysis appears in Ref. [1].

An example of a silicon retina is introduced here. The silicon retina is a type of smart CMOS image sensor using a resistive network architecture [42]. The resistor here is realized by a MOSFET [171]. Like edge detection processing in a retina, a silicon retina processes edges or zero crossings [175] of the input light pattern. The basic circuit is shown in Fig. 3.6(a). A one-dimensional network is illustrated in Fig. 3.6(b), where the input light-converted voltage $V_i(k)$ is input to the network and diffused. The diffused voltage $V_n(k)$ and the input light signal $V_i(k)$ are input into the differential amplifier, the output of which is $V_o k$. The resistive network acts as a smoother of the input light pattern or blurring , as shown in Fig. 3.6(c). It mimics horizontal cells in the retina. The photoreceptor is implemented by a phototransistor (PTr) with a logarithmic response, shown in the inset of Fig. 3.6(a). Its function is discussed in Sec. 3.2.2. The on- and off-cells are implemented by the differential

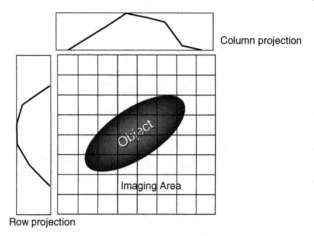

FIGURE 3.5
Projection of an object in the column and row directions.

amplifier. Edge detection is automatically achieved when an image is input. It is noted that two-dimensional resistive networks may be unstable, so careful design is required.

There has been much research into smart CMOS image sensors using resistive networks, as reviewed in Ref. [1]. Recently, a 3T-APS was implemented as a resistive network with noise cancel circuits [45, 46]. Figure 3.7 shows the basic structure of this sensor. It consists of two-layered networks. A 100×100-pixel silicon retina has recently been commercialized for applications in advanced image processing [46]. This sensor processes spatio-temporal patterns so that it can be used, for example, in target tracking, which is discussed in Sec. 4.7.

3.4   Pulse modulation

While in an APS the output signal value is read after a certain time has passed, in pulse modulation (PM) the output signal is produced when the signal reaches a certain value. Such sensors that use PM are called PM sensors, time-to-saturation sensors [176], and address event representation sensors [54]. The basic structure of pulse width modulation (PWM) and pulse frequency modulation (PFM) are shown in Fig. 3.8. Other pulse schemes such as pulse amplitude modulation and pulse phase modulation are rarely used in smart CMOS image sensors. The concept of a PFM-based photosensor was first proposed by K.P. Frohmader [177] and its application to image sensing was first reported by K. Tanaka et al. [108], where a GaAs

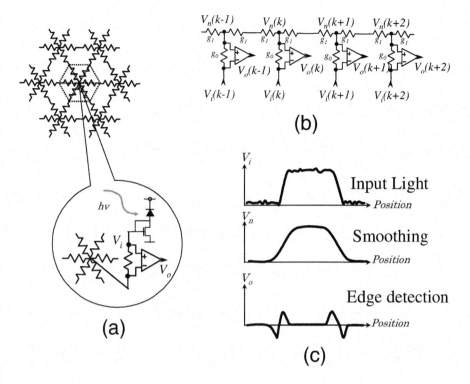

FIGURE 3.6
Illustration of the concept of the silicon retina using a resistive network. (a) Illustration of the circuits, (b) one-dimensional network, (c) input light pattern and its processed patterns [42].

MSM photodetector was used to demonstrate the fundamental operation of the sensor. MSM photodetectors are discussed in Sec. 2.3.5. A PWM-based photosensor was first proposed by R. Müller [178] and its application to an image sensor was first demonstrated by V. Brajovic and T. Kanade [179].

PM has the following features:

- Asynchronous operation

- Digital output

- Low voltage operation

Because each pixel in a PM sensor can individually make a decision to output, a PM sensor can operate without a clock, that is, asynchronously. This feature provides adaptive characteristics for ambient illuminance with a PM-based image sensor and thereby allows application to wide dynamic range image sensors.

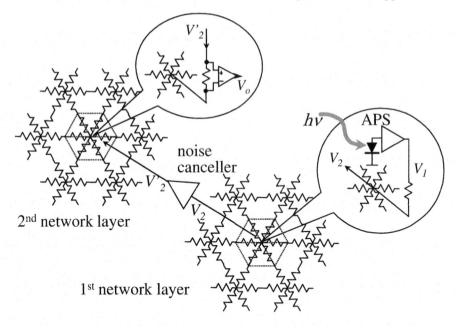

FIGURE 3.7
Architecture of an analog processing image sensor with a two-layered resistive network [45].

Another important feature is that a PM sensor acts as an ADC. In PWM, the count value of the pulse width is a digital value. An example of a PWM is shown in Fig. 3.8, which is essentially equivalent to a single slope type ADC. PFM is equivalent to a one-bit ADC.

In a PM-based sensor, the output is digital, so that it is suitable for low voltage operation. In the next sections, some example PM image sensors are described.

3.4.1 Pulse width modulation

R. Müller first proposed and demonstrated an image sensor based on PWM [178]. Subsequently, V. Brajovic and T. Kanade proposed and demonstrated an image sensor using a PWM-based photosensor [179]. In the sensor, circuits are added to calculate a global operation summing the number of on-state pixels, allowing cumulative evolution to be obtained in an intensity histogram.

The digital output scheme is suitable for on-chip signal processing. M. Nagata et al. have proposed and demonstrated a time-domain processing scheme using PWM and shown that PWM is applicable to low voltage and low power design in deep submicron technology [180]. They have also demonstrated a PWM-based image sensor that realizes on-chip signal processing of block averaging and 2D-projection [181].

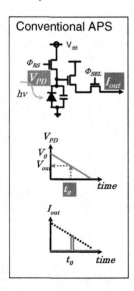
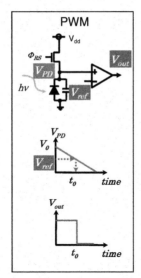
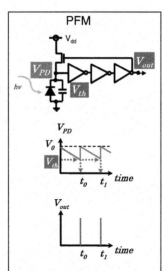

FIGURE 3.8

Basic circuits of pulse modulation. Left: conventional 3T-APS. Middle: pulse width modulation (PWM). Right: pulse frequency modulation (PFM).

The low voltage operation feature of PWM has been demonstrated in Refs. [182, 183], where a PWM based image sensor was operated under a 1-V power supply voltage. In particular, S. Shishido et al. [183] have demonstrated a PWM-based image sensor with pixels consisting of three transistors plus a PD. This design overcomes the disadvantage of conventional PWM-based image sensors that they require a number of transistors for a comparator.

PWM can be applied to enhance a sensor's dynamic range, as described in Sec. 4.4.5, and much research on this topic has been published. Several advantages of this use of PMW, including the improved DR and SNR of PWM, are discussed in Ref. [176].

PWM is also used as a pixel-level ADC in digital image sensors [57, 60–62, 67, 140, 184–186]. Some sensors use a simple inverter as a comparator to minimize the area of a pixel so that a processing element can be employed in the pixels [57, 67]. W. Bidermann *et al.* have implemented a conventional comparator and memory in a chip [186].

In Fig. 3.9(b), a ramp waveform is input to the comparator reference terminal. The circuit is almost the same as a single slope ADC. This type of PWM operates synchronously with the ramp waveform.

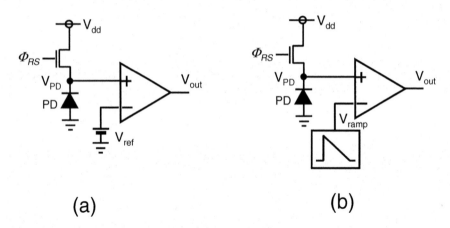

FIGURE 3.9
Basic circuits of pulse width modulation (PWM)-based photosensors . Two types
of PWM photosensor are illustrated: one uses a fixed threshold for a comparator (a)
and the other uses a ramp waveform for a comparator (b).

3.4.2 Pulse frequency modulation

PWM produces an output signal when the accumulation signal reaches a threshold
value. In PFM, when the accumulation signal reaches the threshold value, the output
signal is produced, the accumulated charges are reset and accumulation starts again.
Repeating this process, the output signals continue to be produced. The frequency
of the output signal production is proportional to the input light intensity. PFM-
like coding systems are found in biological systems [187], which have inspired the
pulsed signal processing [188, 189]. K. Kagawa et al. have also developed pulsed
image processing [190], described in Chapter 5. T. Hammadou has discussed s-
tochastic arithmetic in PFM [191]. A PFM-based photosensor was first proposed
and demonstrated by K.P. Frohmader et al. [177]. A PFM-based image sensor was
first proposed by K. Tanaka et al. [108] and demonstrated by W. Yang [192] for a
wide dynamic range, with further details given in Refs. [141, 193, 194].

One application of PFM is address event representation (AER) [195, 196], which
is applied, for example, in sensor network camera systems [197, 198].

PFM-based photosensors are used in biomedical applications, such as in ultra
low light detection in biotechnology [199, 200]. Another application of PFM in the
biomedical field is retinal prosthesis. The application of PFM photosensors to the
retinal prosthesis of subretinal implantation was first proposed in Ref. [201] and has
been continuously developed by the same group [190, 193, 202–212] and by other
groups [213–216]. The retinal prosthesis is described in Sec. 5.4.2.

3.4.2.1   Operation principle of PFM

The operation principle of PFM is as follows. Figure 3.10 shows a basic circuit of a PFM photosensor cell. From the circuit, the sum of the photocurrent I_{ph} including the dark current I_d discharges the PD capacitance C_{PD}, which is charged to V_{dd}, causing V_{PD} to decrease. When V_{PD} reaches the threshold voltage V_{th} of the inverter, the inverter chain is turned on and an output pulse is produced. The output frequency f is approximately expressed as

$$f \approx \frac{I_{ph}}{C_{PD}(V_{dd} - V_{th})}. \tag{3.13}$$

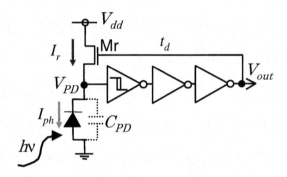

FIGURE 3.10
Basic circuits of a PFM photosensor.

Figure 3.11 shows experimental results for the PFM photosensor in Fig. 3.10. The output frequency increases in proportion to the input light intensity. The dynamic range is measured to be near 100 dB. In the low light intensity region, the frequency saturates due to the dark current.

The inverter chain including the Schmitt trigger has a delay of t_d and the reset current I_r provided by the reset transistor M_r has a finite value. The following is an analysis considering these parameters; a more detailed analysis appears in Ref. [206]. The PD is discharged by the photocurrent I_{ph}; it is charged by the reset current I_r minus I_{ph}, because a photocurrent is still generated during charging.

By taking t_d and I_r into consideration, V_{PD} varies as in Fig. 3.12. From the figure, the maximum voltage V_{max} and minimum voltage V_{min} at V_{PD} are expressed as

$$V_{max} = V_{thH} + \frac{t_d(I_r - I_{ph})}{C_{PD}}, \tag{3.14}$$

$$V_{min} = V_{thL} - \frac{t_d I_{ph}}{C_{PD}}. \tag{3.15}$$

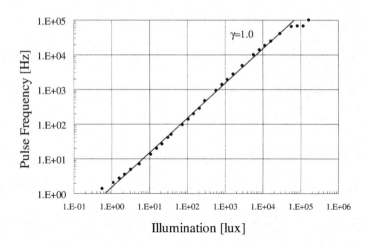

FIGURE 3.11
Experimental output pulse frequency dependence on the input light intensity for the PFM photosensor in Fig. 3.10 .

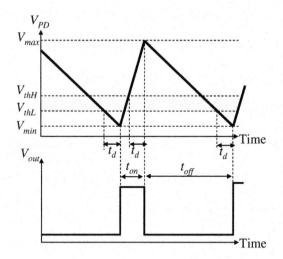

FIGURE 3.12
Time course of V_{PD}, taking the delay t_d into consideration.

Here, V_{thH} and V_{thL} are the upper and lower thresholds of the Schmidt trigger. It is noted that the discharging current is I_{ph}, while the charging current or reset current is $I_r - I_{ph}$. t_{on} and t_{off} in Fig. 3.12 are given by

$$t_{on} = \frac{C_{PD}(V_{thH} - V_{min})}{I_r - I_{ph}} + t_d$$
$$= \frac{C_{PD}V_{th} + t_d I_r}{I_r - I_{ph}},$$

(3.16)

$$t_{off} = \frac{C_{PD}(V_{max} - V_{thL})}{I_{ph}} + t_d$$
$$= \frac{C_{PD}V_{th} + t_d I_r}{I_{ph}},$$

(3.17)

where $V_{th} = V_{thH} - V_{thL}$. t_{on} is the time when the reset transistor Mr charges the PD, that is, when Mr turns on. During this time, the pulse is on-state and hence it is equal to the pulse width. t_{off} is the time when Mr turns off. During this time the pulse is off-state. The pulse frequency f of the PFM photosensor is expressed as

$$f = \frac{1}{t_{on} + t_{off}}$$
$$= \frac{I_{ph}(I_r - I_{ph})}{I_r(C_{PD}V_{th} + t_d I_r)}$$
$$= \frac{I_r^2/4 - \left(I_{ph} - I_r/2\right)^2}{I_r(C_{PD}V_{th} + t_d I_r)}.$$

(3.18)

If the reset current of Mr I_r is much larger than the photocurrent I_{ph}, then Eq. 3.18 becomes

$$f \approx \frac{I_{ph}}{C_{PD}V_{th} + t_d I_r}.$$

(3.19)

Thus the pulse frequency f is proportional to the photocurrent I_{ph}, that is, the input light intensity.

In addition, Eq. 3.18 shows that the frequency f becomes maximum at a photocurrent of $I_r/2$, and then decreases. Its maximum frequency f_{max} is

$$f_{max} = \frac{I_r}{4(C_{PD} + t_d I_r)}.$$

(3.20)

The pulse width τ is

$$\tau = t_{on} = \frac{C_{PD}V_t h + t_d I_r}{I_r - I_{ph}}.$$

(3.21)

From Eq. 3.21, it is seen that if the reset current is equal to the photocurrent, the pulse width becomes infinite, that is, when the input light intensity is strong, or the reset current is small, the pulse width is broadened. Figure 3.14 shows experimental results for the circuits in Fig. 3.10. The pulse width is broadened when V_{dd} is 0.7 V and the input light intensity is large. The reset current depends on the power supply voltage V_{dd} and thus the result of the pulse width broadening is reasonable.

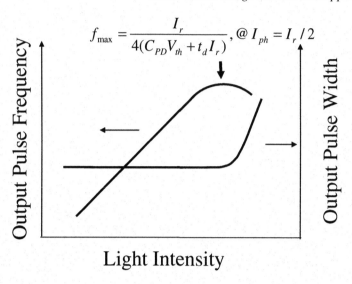

FIGURE 3.13

Pulse frequency and pulse width dependence on the input light intensity.

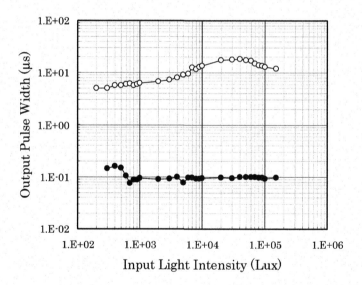

FIGURE 3.14

Experimental results of pulse width dependence on input light intensity for the circuits in Fig. 3.10.

3.4.2.2 Capacitive feedback PFM

The above competition effect between the reset current and photocurrent can be alleviated by introducing capacitive feedback [54, 193]. Fig. 3.15 shows the schematics of an improved circuit with capacitive feedback. In a capacitive feedback PFM photosensor, when V_{pd} reaches a value close to the threshold voltage of INV1, the output of INV2 gradually changes from the LO state to the HI state. This output voltage is positively fedback to V_{pd} through an overlap capacitance C_{rst} of the reset transistor M_{rst} and speeds up the decrease of V_{pd}. The positive feedback action brings about several advantages, most notably that the competition between the reset current and photocurrent is greatly reduced, and thus the power supply voltage can be decreased without any degradation of the pulse characteristics. In addition, since the positive feedback does not require any delay time, produced by an inverter chain, the number of inverters can be reduced. In our experiments, only two stages of inverters were sufficient for satisfactory operation. Experimental results using a capacitive feedback photosensor fabricated in 0.35-μm standard CMOS technology are shown in Fig. 3.16 [209]. Even under a power supply voltage of 1 V, the operation of the photosensor is satisfactory.

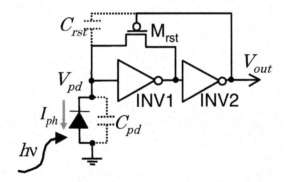

FIGURE 3.15
Schematic of a PFM pixel with capacitive feedback reset [193].

3.4.2.3 PFM with constant PD bias

The above architecture of PFMs changes the PD voltage. It changes linearly according to the input light intensity or photocurrent. Figure 3.17 shows a PFM pixel circuit with a constant PD bias [199]. The node voltage of the PD cathode is a virtual node of an operational amplifier OPamp, so that the node voltage is fixed at V_{ref}. The stored charge of the feedback capacitor C_{int} is extracted by a photocurrent and thus the node voltage at the output of the OPamp increases. When the output of the OPamp reaches the threshold of the comparator Comp, Comp turns on and the reset

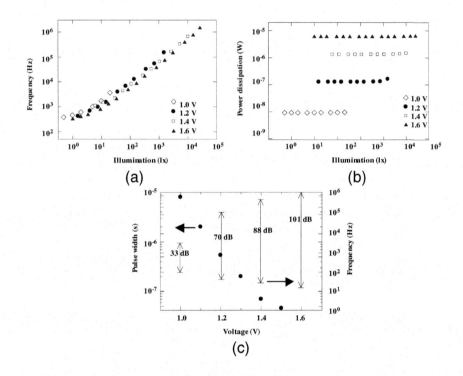

FIGURE 3.16
Experimental results of the dependence of (a) output pulse frequency, (b) power dissipation on the illumination, and (c) the dependence of pulse width and saturation frequency on the power supply voltage of a capacitive feedback PFM photosensor. The photosensor is fabricated in 0.35-μm standard CMOS technology with a pixel size of 100×100 μm^2 and photodiode size of 7.75×15.60 μm^2 [209].

transistor M_{rst} is turns off by the output pulse from Comp.

The frequency of the output pulse from Comp is proportional to the input light intensity. The most important feature of this PFM pixel circuit is that the bias voltage of the PD is constant. This feature is effective in suppressing the dark current of a PD by reducing the bias voltage of the PD. It used for low light detection, discussed in Sec. 4.2.

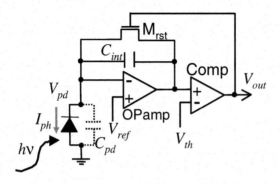

FIGURE 3.17

PFM pixel circuit with constant PD bias [199]. OPamp: operational amplifier, Comp: comparator.

3.5   Digital processing

Digital processing architecture in a smart CMOS image sensor is based on the concept of employing a digital processing element in each pixel [44, 57, 58, 61, 62, 67], Employing digital processing elements in a pixel enables programmable operation with fast processing speed. Also, the nearest neighboring operation can be achieved using digital architecture. Figure 3.18 shows a pixel block diagram reproduced from Ref. [63].

The key point of digital processing architecture is implementing an ADC in a pixel. In Ref. [44], a threshold is introduced for binary operation. In Ref. [57, 67], a simple PWM scheme is introduced with an inverter for a comparator. It is noted that this sensor requires no scanning circuits, because each pixel transfers digital data to the next pixel. This is another feature of fully programmable digital processing architecture. In Ref. [217], a pixel-level ADC is effectively used to control the conversion curve or gamma value. The programmability of the sensor is utilized to enhance the captured image, as in logarithmic conversion, histogram equalization, and similar techniques.

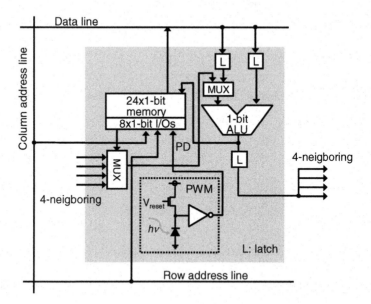

FIGURE 3.18
Pixel circuit diagram of a smart CMOS image sensor with digital processing architecture [63].

Full digital processing architecture is very attractive for smart CMOS image sensors because it is programmable with a definite precision. Thus it is suitable for robot vision, which requires versatile and autonomous operation with a fast response. The challenge for digital processing technology is pixel resolution, which is currently restricted by the large number of transistors in a pixel; for example, in Ref. [61], 84 transistors are integrated in each pixel with a size of 80 μm $\times$ 80 μm using 0.5-μm standard CMOS technology. By using finer CMOS technology, it would be possible to make a smaller pixel with lower power consumption and higher processing speed. However, such fine technology has problems with low voltage swing, low photosensitivity, etc.

Another type of digital processing in smart CMOS image sensors is the digital pixel sensor where an ADC is shared by four pixels [140, 185, 186]. Digital pixel sensors have achieved high speeds of 10000 fps [185] and a wide dynamic range of over 100 dB [186].

3.6 Materials other than silicon

In this section, several materials other than silicon are introduced for smart CMOS image sensors. As described in Chapter 2, visible light is absorbed in silicon, which means that silicon is opaque to visible wavelengths. Some materials are transparent in the visible wavelength region, such as SiO_2 and sapphire (Al_2O_3), which are introduced in modern CMOS technology as SOI (silicon-on-insulator) and SOS (silicon-on-sapphire). The detectable wavelength of silicon is determined by its bandgap, which corresponds to a wavelength of about 1.1 μm. Other materials, such as SiGe and germanium can respond to longer wavelength light than silicon, as summarized in Appendix A.

3.6.1 Silicon-on-insulator

Recently, SOI CMOS technology has been developed for low voltage circuits [218]. The structure of SOI is shown in Fig. 3.19, where a thin Si layer is placed on a buried oxide (BOX) layer. The top Si layer is placed on a SiO_2 layer or an insulator. A conventional CMOS transistor is called a bulk MOS transistor to clearly distinguish it from an SOI MOS transistor. MOS transistors are fabricated on an SOI layer and are completely isolated via shallow trench isolation (STI), which penetrates to the BOX layer, as shown in Fig. 3.19(b). These transistors exhibit lower power consumption, less latch up, and less parasitic capacitance compared with bulk CMOS technology, as well as other advantages [218].

SOI technology is attractive for CMOS image sensors for the following reasons.

- SOI technology can produce circuits with low voltage and low power consumption [219]. For mobile applications, sensor networks, and implantable

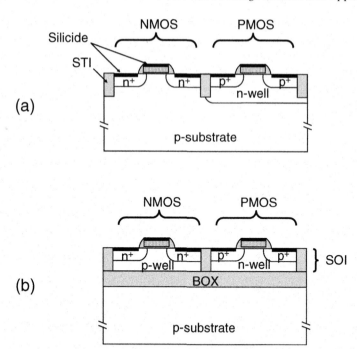

FIGURE 3.19
Cross-section of (a) bulk and (b) SOI CMOS devices. STI: shallow trench isolation, SOI: silicon on insulator, BOX: buried oxide. Silicide is a compound material with silicon and metal such as $TiSi_2$.

medical devices, this feature is important.

- When using SOI processes, NMOS and PMOS transistors can be employed without sacrificing area, compared with bulk CMOS processes where a Newell layer is necessary to construct PMOS transistors on a p-type substrate. Figure 3.19, comparing bulk and SOI, clearly illustrates this. A PMOSFET reset transistor is preferable for use in an APS due to the fact that it exhibits no voltage drop, in contrast with an NMOSFET reset transistor.

- SOI technology makes it easy to fabricate a back-illuminated image sensor, discussed later in this section.

- The SOI structure is useful for preventing crosstalk between pixels [220] caused by diffused carriers; photocarriers generated in a substrate can reach the pixels in the SOI image sensor. In SOI, each pixel is isolated electrically.

- SOI technology can also be used in three-dimensional integration [221, 222]. A pioneering work on the use of SOI in an image sensor for three-dimensional integration appears in Ref. [221].

An issue of applying SOI technology to CMOS image sensors is how to realize photodetection. Generally an SOI layer is so thin (usually below 200 nm) that the photosensitivity is degraded. To obtain good photosensitivity, several methods have been developed. The most compatible method with conventional APSs is to create a PD region on a substrate [223–226]. This ensures that the photosensitivity is the same as that of a conventional PD. However, it requires modifying the standard SOI fabrication process. Post processing for the surface treatment is also important to obtain low dark current. The second method is to employ a lateral PTr, as shown in Fig. 2.7 (d) in Sec. 2.3 [227–230]. As a lateral PTr has a gain, the photosensitivity increases even in a thin photodetection layer. Another application of SOI is in lateral pin PDs [231], although in this case the photodetection area and pixel density is a trade-off. A PFM photosensor, described in Sec. 3.4.2, is especially effective as an SOS imager, which is discussed later in this section.

SOI is widely used in micro-electro-mechanical systems (MEMS) due to the fact that a beam structure of silicon can be easily fabricated by etching a BOX layer with a selective SiO_2 etchant. An application of such a structure in an image sensor is an uncooled focal plane array (FPA) infrared image sensor. [232]. The IR detection is achieved with a thermally isolated pn-junction diode; thermal radiation moves the pn-junction built-in potential and by sensing this shift the temperature can be measured and thus IR radiation is detected. By combining MEMS structures, the range of potential applications of SOI for image sensors will be wide.

3.6.1.1 Back-illuminated image sensor

Here we introduce transparent materials in the visible wavelength region, suitable for back-illuminated image sensors. As shown in Fig. 3.20, a back-illuminated CMOS image sensor has the advantages of a large FF and a large optical response angle. Figure 3.20(a) shows a cross-section of a conventional CMOS image sensor, where the input light travels a long distance from micro-lens to the PD, which causes crosstalk between pixels. In addition, metal wires form obstacles for the light. In a back-illuminated CMOS image sensor, the distance between the micro-lens and the PD can be reduced so that the optical characteristics are much improved. As the p-Si layer on the PD must be thin to reduce absorption in the layer as much as possible, the substrate is typically ground to be thin.

3.6.1.2 Silicon-on-sapphire

Silicon-on-sapphire (SOS) is a technology using sapphire as a substrate instead of silicon [233]. A thin silicon layer is directly formed on a sapphire substrate. It is noted that the top silicon layer is not poly nor amorphous silicon but a single crystal of silicon, and thus the physical properties, such as mobility, are almost the same as in an ordinary Si-MOSFET. Sapphire is Al_2O_3. It is transparent in the visible wavelength region and hence image sensors using SOS technology can be used as back-illuminated sensors without any thinning process [203, 231, 234, 235],

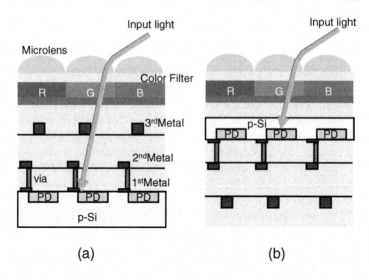

(a)   (b)

FIGURE 3.20
Cross-section of (a) a conventional CMOS image sensor and (b) a back-illuminated
CMOS image sensor.

although some polishing is required to make the back surface flat. Lateral PTrs are
used in the work of Ref. [231, 235], while a PFM photosensor is used in the work of
Ref. [203, 234] due to the low photosensitivity in a thin detection layer. Figure 3.21
shows an image sensor fabricated by SOS CMOS technology. The chip is placed on
a sheet of printed paper and the printed pattern on the paper can be seen through the
transparent substrate.

FIGURE 3.21
PFM photosensor fabricated using SOS technology [203].

3.6.2 Extending the detection wavelength

Usually silicon has a sensitivity of up to 1.1 μm, determined by the bandgap of silicon $E_g(\text{Si}) = 1.12$ eV. To extend the sensitivity beyond 1.1 μm, materials other than silicon must be used. There are many materials with a sensitivity at longer wavelengths than that of silicon. To realize a smart CMOS image sensor with a sensitivity at longer wavelengths than silicon, hybrid integration of materials with a longer wavelength photoresponsivity, such as SiGe, Ge, HgCdTe, InSb, and quantum-well infrared photodetector (QWIP) [236], as well as others [237]. Besides SiGe, these materials can be placed on a silicon readout integrated circuit (ROIC) bonded by flip-chip bonding through metal bumps. Several methods to realize IR detection using ROIC are given in Ref. [238]. Schottky barrier photodetectors such as PtSi (platinum silicide) are also widely used in IR imagers [239], which can be monolithically integrated on a silicon substrate. These infrared image sensors usually work under cooled conditions. There have been many reports of image sensors with this configuration and it is beyond the scope of this book to introduce them. Here we introduce only one example of a smart CMOS image sensor with sensitivities in both the visible and near infrared (NIR) regions, called the eye-safe wavelength region. The human eye is more tolerant to the eye-safe wavelength region (1.4–2.0 μm) than the visible region because more light in the eye-safe region is absorbed at the cornea than light in the visible region and thus less damage is done to the retina.

Before describing the sensor, we briefly overview the materials SiGe and germanium. $\text{Si}_x\text{Ge}_{1-x}$ is a mixed crystal of silicon and germanium with arbitrary composition x [240]. The bandgap can be varied from the bandgap of silicon ($x = 1$), $E_g(\text{Si}) = 1.12$ eV or $\lambda_g(\text{Si}) = 1.1$ μm to that of germanium ($x = 0$), $E_g(\text{Ge}) = 0.66$ eV or $\lambda_g(\text{Ge}) = 1.88$ μm. SiGe on silicon is used for hetero-structure bipolar transistors (HBT) or strained MOS FETs in high-speed circuits. The lattice mismatch between the lattice constants of silicon and germanium is so large that it is difficult to grow thick SiGe epitaxial layers on a silicon substrate.

Now we introduce a smart image sensor that works in both the visible and eye-safe wavelength regions [241, 242]. The sensor consists of a conventional Si-CMOS image sensor and a Ge PD array formed underneath the CMOS image sensor. The capability of the sensor to capture visible images is not effected by extending its range into the IR region. The operating principle of the NIR detection is based on photocarrier injection into a silicon substrate from a Ge PD. The structure of the device is shown in Fig. 3.22. Photo-generated carriers in the germanium PD region are injected into the silicon substrate and reach the photoconversion region at a pixel in the CMOS image sensor by diffusion. When the bias voltage is applied, the responsivity in the NIR region increases, as shown in Fig. 3.23. The inset of the figure shows a test device for the experiment of the photoresponse. It is noted that NIR light can be detected at the Ge PD placed at the back of the sensor because the silicon substrate is transparent in the NIR wavelength region.

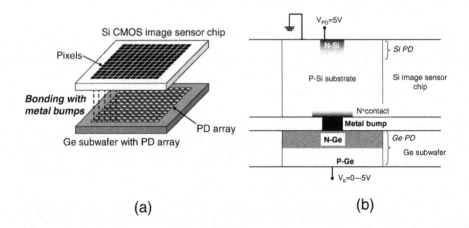

(a)   (b)

FIGURE 3.22

Smart CMOS image sensor that can detect in both the visible and eye-safe regions.
(a) Chip structure, (b) cross-section of the sensor [242].

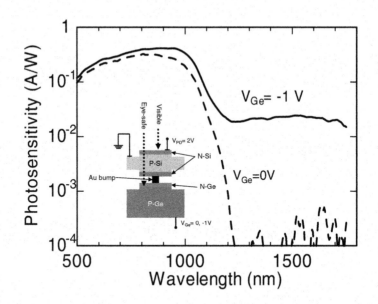

FIGURE 3.23

Photosensitivity curve as a function of input light wavelength. The bias voltage V_b
of the Ge PD is a parameter. The inset shows the test structure for this measurement
[242].

3.7 Structures other than standard CMOS technologies

3.7.1 3D integration

Three-dimensional (3D) integration has been developed to integrate more circuits in a limited area [243, 244]. The interconnections between layers are realized by micro-vias [221, 222, 243, 244], inductive coupling [245, 246], capacitive coupling [247], and optical coupling [248]. Some sensors use SOI [221, 222, 243] and SOS [247] technologies, which make it easy to bond two wafers.

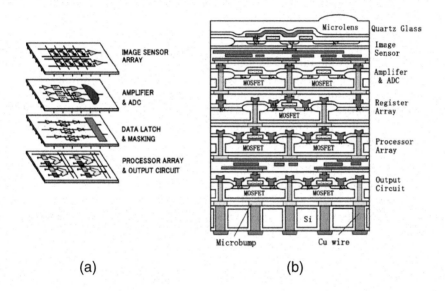

(a)　　　　　　　　　　　　　(b)

FIGURE 3.24

Three-dimensional image sensor chip. (a) Configuration, (b) cross-sectional structure [249].

Since biological eyes have a vertical layered structure, 3D integration can be suitable for mimicing such biological systems [249, 250]. An image sensor with a 3D integration structure has an imaging area on its top surface and signal processing circuits in the successive layers. 3D integration technology thus makes it easy to realize pixel-level processing or pixel-parallel processing. Figure 3.24 shows a conceptual illustration of a 3D image sensor chip and its cross-sectional structure [249]. A 3D image sensor has been proposed to function as a retinal prosthesis device [213], which is discussed in Sec. 5.4.2. 3D image sensors are promising for their capability for pixel-parallel processing, although further development is required to achieve image quality comparable to conventional 2D image sensors.

3.7.2   Integration with light emitters

The integration of light sources in a CMOS image sensor will open many applications, such as ultra small camera systems and autonomous robot vision. III–V compound semiconductors, such as GaAs [86], silicon nano-crystals, and erbium-doped silicon [251], β-FeSi$_2$, emit light with good efficiency, but they are less compatible with standard CMOS technology. In addition, the emission wavelength of erbium-doped silicon and β-FeSi$_2$ is longer than the bandgap wavelength of silicon. Although silicon has an indirect bandgap, it can emit light by band-to-band emission [252] with a fairly good emission efficiency of 1% and by hot electron emission with a lower emission efficiency [253–257]. Band-to-band emission is obtained by a forward bias voltage to a pn-diode. The emission peak is around 1.16 μm, which is determined by the silicon bandgap energy, and thus it is not used for a light source for CMOS image sensors because the photosensitivity at this wavelength is quite low. When a pn-diode is reverse biased, the diode emits light with a broad spectrum (over 200 nm) with a center wavelength of about 700 nm, which can be detected by a silicon PD. This broad emission originates from hot electrons when avalanche breakdown occurs [258, 259].

An image sensor integrated with a Si-LED using a standard SiGe-BiCMOS process technology has been demonstrated [255]. The reason why the SiGe-BiCMOS process is used is that a p^+n^+-diode can be obtained by using the junction between a p^+-base region and a n^+-sinker region, as shown in Fig. 3.25, which are only available for the SiGe-BiCMOS process. It is noted that the emission is not from the SiGe base region, but from the silicon region. Light emission from SiGe-BiCMOS

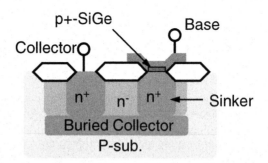

FIGURE 3.25
Cross-section of an LED using SiGe-BiCMOS. A reverse bias voltage is applied between the p^+-base and the n^+-collector [255].

circuits is shown in Fig. 3.26. A CMOS image sensor with an Si-LED array has been fabricated, as shown in Fig. 3.27. Presently, an Si-LED integrated in an image sensor has very high power consumption or low emission efficiency. By optimizing the structure, these characteristics should be improved.

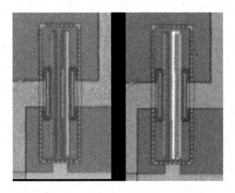

FIGURE 3.26
Emission from an LED using SiGe-BiCMOS technology. Left: without bias. Right: with bias [255].

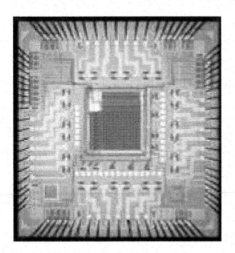

FIGURE 3.27
Image sensor integrated with Si-LEDs [255].

3.7.3   Color realization using nonstandard structures

Usually, image sensors can detect color signals and can separate light into elementary color signals, such as RGB. Conventional methods for color realization are described in Sec. 5.4.1. Other methods to realize color using smart functions are summarized in the next few subsections.

3.7.3.1   Stacked organic PC films

The first introduced method to acquire RGB-color used three stacked photoconductive (PC) organic films that can be fabricated on a pixel [104–106]. Each of the organic films acts a PC detector (see Sec. 2.3.5) and produces photocurrents according to its light sensitivity. This method can almost realize a 100% FF. The main issue is how to connect the stacked layers.

3.7.3.2   Multiple junctions

The photosensitivity in silicon depends on the depth of the pn-junction. Thus, having two or three junctions located along a vertical line alters the photosensitivity spectrum [260–262]. To adjust the three junction depths, the maximum photosensitivities corresponding to the RGB colors are realized. Figure 3.28 shows the structure of such a sensor, where a triple well is located to form three different photodiodes [261, 262]. This sensor has been commercialized as an APS type pixel.

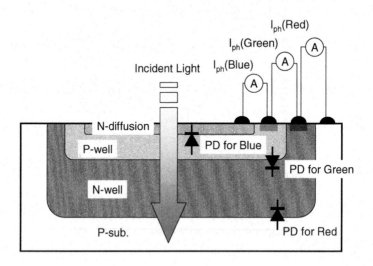

FIGURE 3.28
Device structure of image sensor with a triple junction [262]. In the commercialized sensor, an APS structure is employed [262].

3.7.3.3 Controlling the potential profile

Changing the spectrum sensitivity by controlling the potential profile has been proposed and demonstrated by many researchers [263–265]. The proposed systems mainly use a thin film transistor (TFT) layer consisting of multiple layers of p-i-i-i-n [263, 264] and n-i-p-i-n [265]. Y. Maruyama et al. at Toyohashi Univ. Technology have proposed a smart CMOS image sensor using such a method [266,267], although their aim is not for color realization but for filterless fluorescence detection, which is discussed in Sec. .

The principle of potential control is as follows [266, 267]. As discussed in Sec. 2.3.1.2, the sensitivity of a pn-junction PD is generally expressed by Eq. 2.19. Here we use the potential profile shown in Fig. 3.29. This figure is a variation of Fig. 2.11 by replacing the NMOS-type PG with a PMOS-type PG on an n-type substrate, giving two depletion regions, one originating from the PG and the other from the pn-junction. This pn-junction produces a convex potential that acts like a watershed for photo-generated carriers. In this case, the integral region in Eq. 2.18 is changed

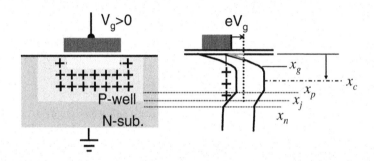

FIGURE 3.29

Device structure and potential profile for the filterless fluorescence image sensor [266, 267].

from 0 to x_c, where photo-generated carriers have an equal chance to flow to the surface or to the substrate. Carriers that flow to the substrate only contribute to the photocurrent. The sensitivity thereby becomes

$$
\begin{aligned}
R_{ph} &= \eta_Q \frac{e\lambda}{hc} \\
&= \frac{e\lambda}{hc} \frac{\int_0^{x_c} \alpha(\lambda) P_o \exp[-\alpha(\lambda)x]\,dx}{\int_0^{\infty} \alpha(\lambda) P_o \exp[-\alpha(\lambda)x]\,dx} \\
&= \frac{e\lambda}{hc} \left(1 - \exp[-\alpha(\lambda)x_c]\right).
\end{aligned}
\tag{3.22}
$$

From this, if two lights with different wavelengths, excitation light *lambda*$_{ex}$ and fluorescence λ_{fl}, are incident simultaneously, then the total photocurrent I_{ph} is given by

$$I_{ph} = P_o(\lambda_{ex})A\frac{e\lambda_{ex}}{hc}\left(1 - \exp[-\alpha(\lambda_{ex})x_c]\right) + P_o(\lambda_{fl})A\frac{e\lambda_{fl}}{hc}\left(1 - \exp\left[-\alpha(\lambda_{fl})x_c\right]\right),$$
(3.23)

where $P_o(\lambda)$ and A are the incident light power density with λ and the photogate (PG) area. When measuring the photocurrent with two different gate voltages, x_c has two different values x_{c1} and x_{c2}, which results in two different photocurrents I_{ph1} and I_{ph2}:

$$I_{ph1} = P_o(\lambda_{ex})A\frac{e\lambda_{ex}}{hc}\left(1 - \exp\left[-\alpha(\lambda_{ex})x_{c1}\right]\right) + P_o(\lambda_{fl})A\frac{e\lambda_{fl}}{hc}\left(1 - \exp\left[-\alpha(\lambda_{fl})x_{c1}\right]\right),$$

$$I_{ph2} = P_o(\lambda_{ex})A\frac{e\lambda_{ex}}{hc}\left(1 - \exp\left[-\alpha(\lambda_{ex})x_{c2}\right]\right) + P_o(\lambda_{fl})A\frac{e\lambda_{fl}}{hc}\left(1 - \exp\left[-\alpha(\lambda_{fl})x_{c2}\right]\right).$$
(3.24)

In these two equations, the unknown parameters are the input light intensities $P_o(\lambda_{ex})$ and $P_o(\lambda_{fl})$. We can calculate the two input light powers, $P_o(\lambda_{ex})$ for the excitation light and $P_o(\lambda_{fl})$ for the fluorescence power, that is, filterless measurement can be achieved.

3.7.3.4   Sub-wavelength structure

The fourth method to realize color detection is to use a sub-wavelength structure such as a metal grid or surface plasmons [268–270] and photonic crystals [271]. These technologies are in their preliminary stages but may be effective for CMOS image sensors with fine pitch pixels. In sub-wavelength structures, the quantum efficiency is very sensitive to polarization as well as the wavelength of the incident light and the shape and material of the metal grid. This means that the light must be treated as an electro-magnetic wave to estimate the quantum efficiency.

When the diameter of an aperture d is much smaller than the wavelength of the incident light λ, the optical transmission through the aperture T/f, which is the transmitted light intensity T normalized to the intensity of the incident light in the area of the aperture f, decreases according to $(d/\lambda)^4$ [272], which causes the sensitivity of an image sensor to decrease exponentially. T. Thio et al. reported a transmission enhancement through a sub-wavelength aperture surrounded by periodic grooves on a metal surface [273]. In such a structure, surface plasmon (SP) modes are excited by the grating coupling of the incident light [274], and resonant oscillation of SPs causes an enhancement of the optical transmission through the aperture. This transmission enhancement could make it possible to realize an image sensor with a sub-wavelength aperture. From computer simulation results given in Ref. [269], an aluminum metal grid enhances optical transmission, a while tungsten metal grid does not enhance it. The thickness and the line and space of the metal grid also influence the transmission.

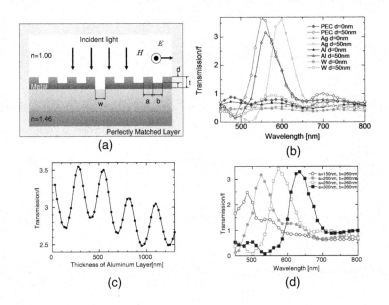

FIGURE 3.30

Metal grid as a sub-wavelength structure and its optical transmission. Computer simulation (FDTD: Finite Differential Time Domain) model (a). Optical transmission dependence on the wavelength with the parameters of materials (b), metal thickness (c) and period (d). PEC: perfect electric conductor [269].

4

Smart imaging

4.1 Introduction

Some applications require imaging which is difficult to achieve by using conventional image sensors, either because of limitations in their fundamental characteristics, such as speed and dynamic range, or because of the need for advanced functions such as target tracking and distance measurement. For example, intelligent transportation systems (ITSs) in the near feature will require intelligent camera systems to perform lane keeping assisting, distance measurement, driver monitor, and other functions, for which smart image sensors must be applied with a wide dynamic range of over 100 dB, high speed over video rate, and the ability to measure distances of multiple objects in an image [275]. Security, surveillance, and robot vision are similar applications to ITSs. Smart imaging is also effective in the information and communication fields as well as in biomedical fields.

Many types of implementation have been developed to integrate the smart functions described in the previous chapter. The functions are commonly classified by the level where the function is processed, pixel level, column level, and chip level. Figure 4.1 shows smart imaging in CMOS image sensors, illustrating the classification. Of course, a mixture of levels in one system is possible. The most straightforward implementation is chip-level processing where signal processing circuits are placed after the signal output, as shown in Fig. 4.1 (a). This is an example of a "camera-on-a-chip," where an ADC, noise reduction system, color signal processing block, and other elements are integrated in a chip. It is noted that this type of processing requires about the same output data rate so that the gain in the signal processing circuits are limited. The second implementation method is column-level processing or column-parallel processing. This is suitable for CMOS image sensors, because the column output lines are electrically independent. As signal processing is achieved in each column, a slower processing speed can be used than for chip-level processing. Another advantage is that the pixel architecture can be the same as in a conventional CMOS image sensor, so that, for example, a 4T-APS can be used. This feature is a great advantage in achieving a good SNR.

The third implementation method is pixel-level processing or pixel-parallel processing. In this method, each pixel has a signal processing circuit as well as a photodetector. This can realize fast, versatile signal processing, although the photodetection area or fill factor is reduced so that the image quality may be degraded

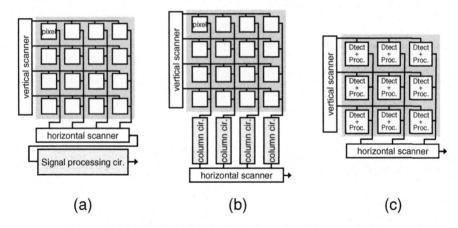

(a)   (b)   (c)

FIGURE 4.1

Basic concept of smart CMOS image sensors. (a) Chip-level processing, (b) column-level processing, and (c) pixel-level processing.

compared with the former two implementation methods. Also, in this method, it is difficult to employ a 4T-APS for a pixel. However, this architecture is attractive and is a candidate for the next generation of smart CMOS image sensors.

Other features of CMOS image sensors for smart functions, used in several applications, are as follows.

- Random access to an arbitrary pixel.

- Non-destructive readout or multiple readout. Note that the original 4T-APS cannot be applied for non-destructive readout.

- Integration of signal processing circuitry in each pixel and/or each column and/or a chip.

In this chapter, we survey smart imaging required for smart CMOS image sensors for several applications.

4.2 Low light imaging

Low light imaging is essential for several applications, such as in the fields of astronomy and biotechnology. Some image sensors have ultra high sensitivity, such as super HARP [102] and Impactron [276, 277], but in this section we focus on smart CMOS image sensors for low light imaging. Some applications in low light imaging do not require video rate imaging so that long accumulation times are allowed.

For long exposure times, dark current and flicker noise or $1/f$ noise are dominant; a detailed analysis for suppressing noise in low light imaging using CMOS image sensors is reported in Refs. [199, 200, 278]. To decrease the dark current of a PD, the most effective, straightforward method is cooling. However, in some applications it is difficult to cool the detector.

Here we discuss how to decrease the dark current at room temperature. First of all, pinned PDs (PPDs) or buried PDs (BPDs), as described in Sec. 2.4.3, are effective in decreasing the dark current.

Decreasing the bias voltage at the PD is also effective [78]. As mentioned in Sec. 2.4.3, the tunnel current strongly depends on the bias voltage. Figure 4.2 shows near zero bias circuits, developed as reported in Ref. [279]. As reported, one near zero bias circuit is located in a chip and provides the gate voltage of the reset transistor.

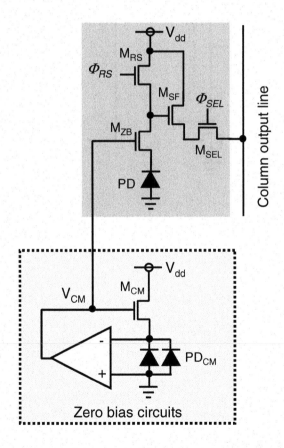

FIGURE 4.2
Near zero bias circuits for reducing dark current in an APS [279].

4.2.1 Active reset for low light imaging

Active reset senses the PD node voltage V_{PD} and stabilizes it by active feedback during the reset phase [280–288]. Figure 4.3 shows an example of an active reset circuit, although there are many variations of active reset implementation. The output voltage of the pixel is input to the operational amplifier outside the pixel and fedback to the gate of the reset transistor M_{rst} to stabilize the PD node voltage. Is it estimated that active reset can reduce $k_B T C$ noise down to $k_B T / 18 C$ [280].

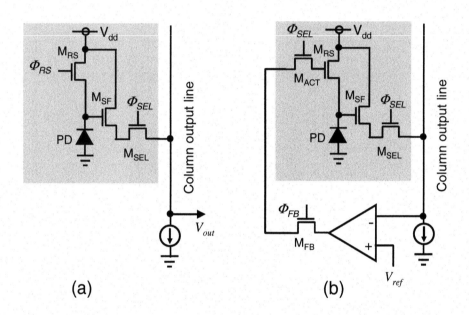

(a)   (b)

FIGURE 4.3
Active reset circuits. (a) Conventional 3T-APS, (b) 3T-APS with active reset [280].

4.2.2 PFM for low light imaging

The PFM photosensor can achieve ultra low light detection with near-zero bias of the PD and can obtain a minimum detectable signal of 0.15 fA (1510 s integration time), as discussed in Sec. 3.4.2.3. As mentioned in Sec. 2.3.1.3, the dark current of a PD is exponentially dependent on the PD bias voltage, so that PD bias near zero voltage is effective in reducing dark current. In such a case, special care must be taken to reduce other leakage currents; in a PFM photosensor with constant PD bias, the leak current or subthreshold current of the reset transistor M_{rst} in Fig. 3.17 is critical and must be reduced as much as possible. Bolton et al. have introduced the circuit shown in Fig. 4.4 where the drain–source voltage of the reset transistor

M_{rst1} is near zero so that the subthreshold current reaches zero. It is noted that the subthreshold current is exponentially dependent on the drain–source voltage as well as the gate–source voltage, as discussed in Appendix E.

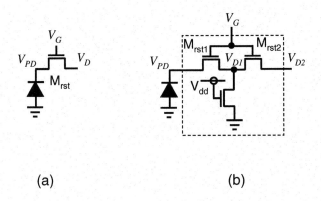

(a)   (b)

FIGURE 4.4

Reset circuit for PFM with constant PD bias. (a) A reset transistor M_{rst} replaces (b) a three transistor circuit [200].

4.2.3   Differential APS

To suppress common mode noise, the differential APS has been developed, as report-ed in Ref. [278], shown in Fig. 4.5, where the sensor uses a pinned PD and low bias operation using a PMOS source follower. It is noted that the amount of $1/f$ noise in a PMOS is less than in an NMOS. Consequently, the sensor demonstrates ultra low light detection of 10^{-6} lux over 30 s of integration time at room temperature.

4.2.4   Geiger mode APD for a smart CMOS image sensor

To achieve ultra low light detection with a fast response, the use of an APD is benefi-cial. Figure 4.6 shows the pixel structure of an APD in standard CMOS technology. The APD is fabricated in a deep n-well and multiple regions are surrounded by a p^+-guard ring region. In the Geiger mode, the APD (in this case the photodiode is called a single photon avalanche diode (SPAD)) produces a spike-like signal, not an analog output. A pulse shaper with an inverter is used to convert the signal to a digital pulse, as shown in Fig. 4.6. The incident light intensity is proportional to the number of produced pulse counts.

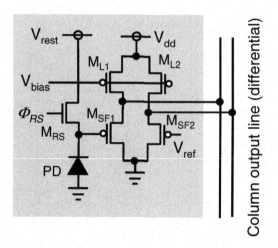

FIGURE 4.5
Differential APS. Illustrated after Ref. [278].

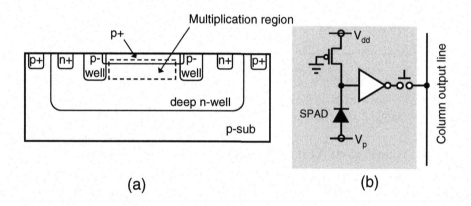

FIGURE 4.6
Basic structure of (a) an APD in standard CMOS technology and (b) a circuit for the Geiger mode APD in a pixel. The PMOS connected to V_{dd} acts as a resistor for quenching process. V_p is a negative voltage value that forces the PD into the avalanche breakdown region. SPAD: single photon avalanche diode. Illustrated after Ref. [98].

4.3 High speed

High speed is an advantageous characteristic for a CMOS image sensor because a column-parallel circuit is suitable for achieving high data rate with a relatively slow processing time in a column, for example, of about 1 μsec. Ultra high-speed cameras have previously been developed based on CCD technology [289], and several types of high-speed cameras based on CMOS image sensors have recently been reported [156, 185, 290–297]. A high speed frame rate alleviates the disadvantage of a rolling shutter inherent to CMOS image sensors.

4.3.1 Global shutter

To acquire a ultra high speed image over 1000 fps, a global shutter is required. Usually, one transistor and a capacitor are added to a 3T-APS pixel to achieve a global shutter function [291, 298, 299]. The pixel circuit is shown in Fig. 4.7. An image lag is critical for ultra high speed images and hence a 4T-APS is not suitable due to its relatively long transfer time. In addition, in 3T-APSs and 5T-APSs, as shown in Fig. 4.7, a hard reset is necessary to ensure there is no image lag [146].

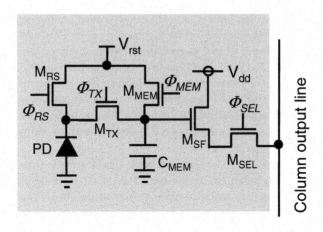

FIGURE 4.7
Basic pixel circuits for global shutter function [291].

4.4   Wide dynamic range

4.4.1   Principle of wide dynamic range

The human eye has a wide dynamic range of about 200 dB. To achieve such a wide dynamic range, the eye has three mechanisms [300]. First, the human eye has two types of photoreceptor cells, cones and rods which correspond to two types of photodiodes with different photosensitivities. Second, the response curve of the eye's photoreceptor is logarithmic so that saturation occurs slowly. Third, the response curve shifts according to the ambient light level or averaged light level. Conventional image sensors, in contrast, have a dynamic range of 60–70 dB, which is mainly determined by the well capacity of the photodiode.

Some applications, such as in automobiles and for security, require a dynamic range over 100 dB [275]. To expand the dynamic range, many methods have been proposed and demonstrated. They can be classified into three categories: nonlinear response, multiple sampling, and saturation detection. Figure 4.8 illustrates these methods. An example image taken by a wide dynamic range image sensor developed by S. Kawahito and his colleagues at Shizuoka Univ. [301] is shown in Fig. 4.9. The extremely bright lightbulb at the left can be seen, as well as the objects under a dark condition to the right.

As mentioned above, the human retina has two types of photoreceptors with high and low sensitivities. An image sensor with two types of PDs with high and low sensitivities can achieve a wide dynamic range. Such an image sensor has already been produced using CCDs. A CMOS image sensor has been reported in Ref. [302] that also has two types of photodetectors with high and low sensitivities, but an FD is used as the photodetector with low sensitivity. The FD is optically shielded but can still gather photo-generated charges under a high illumination condition. Under this operation principle, this sensor achieves a 110-dB intra-scene dynamic range with in CIF.

Nonlinear response is a method to modify the photoresponse from linear to nonlinear, for example, a logarithmic response. This method can be divided into two methods, using a log sensor and well-capacity adjustment. In a log sensor, a photodiode has a logarithmic response. By adjusting the well capacity, the response can be changed to be nonlinear, but in some cases, a linear response is achieved, which is mentioned later.

Multiple sampling is a method where the signal charges are read several times. For example, bright and dim images are obtained with different exposure times and then the two images are synthesized so that both scenes can be displayed in one image. Extending the dynamic range by well capacity adjustment and multiple sampling is analyzed in detail in Ref. [303].

In the saturation detection method the integration signal or accumulation charge signal is observed and if the signal reaches a threshold value, then, for example, the accumulation charge is reset and the reset number is counted. Repeating this pro-

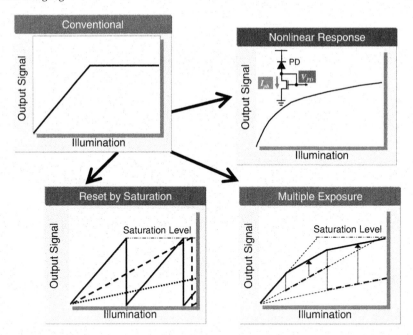

FIGURE 4.8
Basic concept of enhancing dynamic range.

cess, the final output signal is obtained for the residue charge signal and the reset count number. There are several variations to the saturation detection method. Pulse modulation is one alternative and is discussed in Sec. 3.4. In this method, the integration time is different from pixel to pixel. For example, in PWM, the output is the pulse width or counting values so that the maximum detectable light intensity is determined by the minimum countable value or clock and the minimum detectable value is determined by the dark current. Thus the method is not limited by the well capacity, so that it has a wide dynamic range.

The group in Fraunhofer Institute has developed a wide dynamic range image sensor based on local brightness adaptation using a resistive network [304, 305]. Resistive networks are discussed in Sec. 3.3.3. They used a method by which a network can more strongly diffuse a signal at a brighter spot.

In the following sections, examples of the above four methods are described.

4.4.2 Dual sensitivity

When two types of photodetectors with different sensitivities are integrated in a pixel, a wide range of illumination can be covered; under bright conditions, the PD with lower sensitivity is used, while under dark conditions, the PD with higher sensitivity is used. This is very similar to the human visual system, as mentioned above, and

FIGURE 4.9
Example image taken by a wide dynamic range image sensor developed by S. Kawahito et al. [301]. The image is produced by synthesizing several images; the details are given in Sec. 4.4.4. Courtesy of Prof. Kawahito at Shizuoka Univ.

has already been implemented using CCDs [336]. In the work reported in Ref. [302], an FD is used as a low sensitivity photodetector. Under bright light conditions, some carriers are generated in the substrate and diffuse to the FD region, contributing to the signal. This structure was first reported in Ref. [306]. This is a direct method of detection so that no latency of captured images occurs, in contrast with the multiple sampling method. Another implementation of dual photodetectors has been reported in Ref. [307], where a PG is used as the primary PD and an n-type diffusion layer is used as the second PD.

4.4.3 Nonlinear response

4.4.3.1 Log sensor

The log sensor is used for wide dynamic range imagers, and can now obtain a range of over 100 dB [163–169].

Issues of the log sensor include variations of the fabrication process parameters, a relatively large noise in low light intensity, and image lag. These disadvantages are mainly exhibited in subthreshold operation, where the diffusion current is dominant.

There have been some reports of the achievement of logarithmic and linear responses in one sensor [308, 310]. The linear response is preferable in the dark light region, while the logarithmic response is suitable in the bright light region. In these

TABLE 4.1

Smart CMOS sensors for a wide dynamic range

Category	Implementation method	Refs.
Dual sensitivity	PG & PD in a pixel	[302, 306, 307]
Nonlinear response	Log sensor	[163–169]
	Log/linear response	[308–310]
	Well capacity adjustment	[298, 311–318]
	Control integration time/gain	[319, 320]
Multiple sampling	Dual sampling	[321–323]
	Multiple sampling with fixed short exposure time	[324, 325]
	Multiple sampling with varying short exposure time	[301]
	Multiple sampling with pixel-level ADC	[140, 184, 186]
Saturation detection	Locally integration time and gain	[326]
	Saturation count	[263, 326–332]
	Pulse width modulation	[179, 263, 333]
	Pulse frequency modulation	[141, 194, 196, 202, 334, 335]
Diffusive brightness	Resistive network	[304, 305]

sensors, calibration in the transition region is essential.

4.4.3.2 Well-capacity adjustment

The well-capacity adjustment is a method to control the well depth in the charge accumulation region during integration. In this method, a drain for the overflow charges is used. Controlling the gate between the accumulation region and the overflow drain, the photoresponse curve becomes nonlinear.

Figure 4.10 illustrates the pixel structure with an overflow drain to enhance the maximum well capacity. When strong illumination is incident on the sensor, the photo-generated carriers are saturated in the PD well and flow over into the FD node. By decreasing the potential well of the overflow drain (OFD) gradually, strong light intensity can hardly saturate the well and also weak light can be detected.

This method realizes a saturated response with almost the same pixel structure as 3T- and 4T-APSs, and hence it should have a good SNR under low light conditions. The drawback of this method is that the overflow mechanism consumes the pixel area so that the fill factor is reduced.

The method can be implemented in both 3T-APSs [311, 312, 319], and 4T-APSs [298, 313–316]. The sensitivity in a 4T-APS is better than in a 3T-APS and therefore the dynamic range from the dark light condition to the bright light condition can be

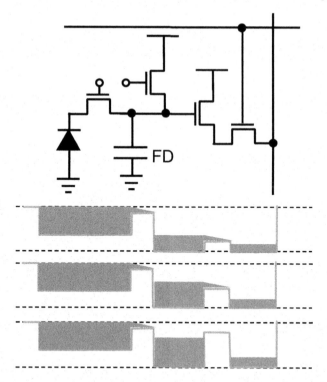

FIGURE 4.10
Pixel structure of an overflow drain type wide dynamic range image sensor [311].

improved.

In Ref. [314], noise reduction by CDS is implemented so that a high SN ratio with 0.15-mV$_{rms}$ of random noise and 0.15-mV$_{rms}$ of FPN are obtained. In this study, a stacked capacitor used for the lateral OFD (overflow drain) was developed. By introducing a direct photocurrent output mode, an ultra high dynamic range of over 200 dB has been achieved, as reported in Ref. [315], by combining with the architecture reported in Ref. [314]. In the region of the direct photocurrent output mode, a logarithmic response is employed.

M. Ikebe at Hokkaido Univ. has proposed and demonstrated a method of PD capacitance modulation using negative feedback resetting [317]. This method does not modify the pixel structure, which is a 3T- or 4T-APS, but instead controls the reset voltage through a column differential amplifier. This method is applied to noise suppression as well as achieving a wide dynamic range.

In Ref. [318], a 3T-APS and a PPS are combined in a pixel to enhance the dynamic range. Generally, an APS has a superior SNR at low illumination compared with a PPS, and this suggests that at bright illumination a PPS is acceptable. A PPS is suitable for use with an OFD because a column charge amplifier can completely transfer charges in both the PD and OFD. In this case, care need not be taken with regard to whether or not signal charges in PD are transferred into the OFD.

4.4.4 Multiple sampling

Multiple sampling is a method to read signal charges several times and synthesize those images in one image. This method is simple and easily achieves a wide dynamic range. However, it has an issue regarding synthesis of the images obtained.

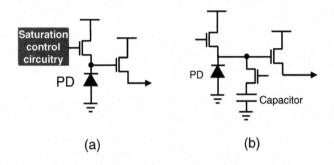

(a)   (b)

FIGURE 4.11
Multiple readout scheme for wide dynamic range. (a) Method to control the reset transistor,(b) method to use another accumulation capacitor.

4.4.4.1 Dual sampling

In dual sampling, two sets of readout circuits are employed in a chip [321, 323]. When pixel data in the n-th row are sampled and held in one readout circuitry and then reset, pixel data in the $(n - \Delta)$-th row are sampled and held in another readout circuitry and reset. The array size is $N \times M$, where N is the number of rows and M is the number of columns. In this case the integration time of the first readout row is $T_l = (N - \Delta)T_{row}$ and the integration time of the second readout row is $T_s = \Delta T_{row}$. Here, T_{row} is the time required to readout one row of data and $T_{row} = T_{SH} + MT_{scan}$, where T_{SH} is a sample and hold (S&H) time and T_{scan} is the time to read out data from the S&H capacitors or the scanning time. It is noted that by using the frame time $T_f = NT_{row}$, T_l is expressed as

$$T_l = T_f - T_s. \tag{4.1}$$

The dynamic range is given by

$$DR = 20\log \frac{Q_{max}T_l}{Q_{min}T_s} = DR_{org} + 20\log \left(\frac{T_f}{T_s} - 1 \right). \tag{4.2}$$

Here, DR_{org} is the dynamic range without dual sampling and Q_{max} and Q_{min} are the maximum and minimum accumulation charges, respectively. For example, if $N = 480$ and $\Delta = 2$, then the ratio of the accumulation $T_f/T_s \approx T_l/T_s$ becomes about 240, so that it can expand the dynamic range of about 47 dB.

This method only requires two S&H regions and no changes of pixel structure, therefore it can be applied, for example, to a 4T-APS, which has high sensitivity. The disadvantages of this method are that only two accumulation times are obtained and a relatively large SNR dip is exhibited at the boundary of the two different exposures.

At the boundary of the two different exposures, the accumulated signal charges are changed from its maximum value Q_{max} to $Q_{max}T_s/T_l$, which causes a large SNR dip. The SNR dip ΔSNR is

$$\Delta SNR = 10\log \frac{T_s}{T_l}. \tag{4.3}$$

If the noise level is not changed in the two regions, ΔSNR is equal to ≈ -24 dB for the above example ($T_l/T_s \approx 240$).

4.4.4.2   Multiple sampling

Fixed short time exposure   To reduce the SNR dip, M. Sasaki et al. introduced multiple short time exposure [324, 325]. In non-destructive readout, multiple sampling is possible. By reading the short time integration T_s a total of k times, the SNR dip becomes

$$\Delta SNR = 10\log k\frac{T_s}{T_l}. \tag{4.4}$$

It is noted that in this case T_l is expressed as

$$T_l = T_f - kT_s. \tag{4.5}$$

Thus the dynamic range expansion ΔDR is

$$\Delta DR = 20\log \left(\frac{T_f}{T_s} - k \right), \tag{4.6}$$

which is little changed from T_f/T_s. If $T_f/T_s = 240$ and $k = 8$, then $\Delta DR \approx 47$ dB and $\Delta SNR \approx -15$ dB.

Varying short exposure times   In the previous method, a short exposure time was fixed and read several times. M. Mase et al. have improved the method by varying the short exposure time [301]. In the short exposure time slot, several different exposure times are employed. During the readout time of one short exposure period, a shorter exposure period is inserted, while a further shorter exposure period is inserted and so on. This is illustrated in Fig. 4.12. With a fast readout circuitry with column parallel cyclic ADCs it is possible to realize this method. In this method, the dynamic range expansion ΔDR is

$$\Delta DR = 20\log \frac{T_l}{T_{s,min}}, \tag{4.7}$$

where $T_{s,min}$ means the minimum exposure time. In this method, the SNR dip can be reduced by making each exposure time the ratio of T_l to Ts to be a minimum.

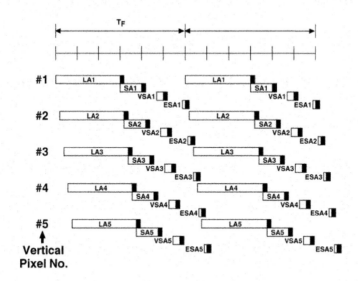

FIGURE 4.12
Exposure and readout timing for multiple sampling with varying short exposure times [301]. LA: long accumulation, SA: short accumulation, VSA: very short accumulation,d ESA: extremely short accumulation.

Pixel-level ADC Another multiple sampling method is to implement pixel-level ADCs [140, 184, 186]. In this method, a bit-serial ADC with single slope is employed for every four pixels. According to the integration time, the precision of the ADC is varied to obtain a high resolution; for a short integration time, a higher precision ADC is executed. The image sensor reported in Ref. [186] is integrated with a DRAM frame buffer in the chip and achieves a dynamic range over 100 dB with a pixel number of 742×554.

4.4.5 Saturation detection

The saturation detection method is based on monitoring and controlling the saturation signal. The method is asynchronized so that automatic exposure of each pixel is easily achieved. A common issue with this method is how to suppress the reset noise; it is difficult to employ a noise cancellation mechanism due to the multiple reset action. A decreased SNR in the reset is also an issue.

4.4.5.1 Saturation count

When the signal level reaches the saturation level, the accumulation region is reset and accumulation starts again. By repeating the process and counting the number of resets in a time duration, the total signal charge in the time period can be calculated with the residue charge signal and the number of resets [263, 326–332]. The counting circuitry is implemented at the pixel level [263, 327–329], and in a column level [326, 330, 331]. At the pixel level, the fill factor is reduced due to the extra area required for the counting circuitry. Using TFT technology, as in Refs. [263, 327], alleviates this issue, although special process technology is required. For either the pixel level or column level, frame memory is required in this method.

4.4.5.2 Pulse width modulation

Pulse width modulation (PWM) is mentioned in Sec. 3.4.1 and is used for wide dynamic range image sensors [179].

4.4.5.3 Pulse frequency modulation

Pulse frequency modulation (PFM) is discussed in Sec. 3.4.2 and is used for wide dynamic range image sensors [141, 192–194].

4.4.6 Diffusive brightness

In this method, the input light is diffused by the resistive network architecture, as mentioned in Sec. 3.3.3 [304, 305]. The output of a bright spot is suppressed via diffusion, so that the photoresponse curve becomes nonlinear. The response speed of the resistive network is not fast and hence it is difficult to apply this method to capturing fast moving objects.

4.5 Demodulation

4.5.1 Principles of demodulation

In the demodulation method, a modulated light signal is illuminated on an object and the reflected light is acquired by the image sensor. The method is effective in detecting signals with a high SNR, because it enables the sensor to detect only modulated signals and thus removes any static background noise. Implementing this technique in an image sensor with a modulated light source is useful for such fields as intelligent transportation systems (ITS), factory automation, and robotics, because such a sensor could acquire an image while being hardly affected by background light conditions. In addition to such applications, this sensor could be applied to tracking a target specified by a modulated light source. For example, motion capture could be easily realized by this sensor under various illumination conditions. Another important application of the demodulation technique is the three-dimensional range finder with the time-of-flight method, as described in Sec. 4.6.

It is difficult to realize demodulation functions in conventional image sensors, because a conventional image sensor operates in accumulation mode so that the modulated signal is washed out by the accumulating modulation charges. The concept

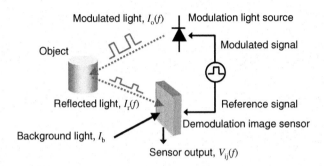

FIGURE 4.13

Concept of a demodulation image sensor.

of the demodulation technique in a smart CMOS image sensor is illustrated in Fig. 4.13. The illumination light $I_o(t)$ is modulated by the frequency f, and the reflected (or scattered) light $I_r(t)$ is also modulated by f. The sensor is illuminated with the reflected light $I_r(t)$ and the background light I_b, that is, the output from the photodetector is proportional to the sum $I_r(t) + I_b$. The output from the photodetector is multiplied by the synchronous modulated signal $m(t)$ and then integrated. Thus the output V_{out} produces [337]

$$V_{out} = \int_{t-T}^{t} \left(I_r(\tau) + I_b \right) m(\tau) d\tau, \qquad (4.8)$$

where T is the integration time.

There are several reports of realizing a demodulation function in smart CMOS image sensors based on the concept shown in Fig. 4.13. Its implementation method can be classified into two categories: the correlation method [166, 337–344], and a method using two accumulation regions in a pixel [345–353]. The correlation method is a straightforward implementation of the concept in Fig. 4.13.

4.5.2 Correlation

The correlation method is based on multiplying the detected signal with the reference signal and then integrating it or performing low-pass-filtering. The process is described by Eq. 4.8. Figure 4.14 shows the concept of the correlation method.

The key component in the correlation method is a multiplier. In Ref. [344], a sim-

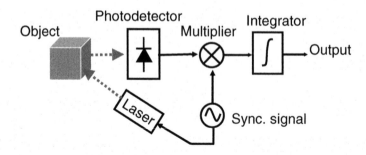

FIGURE 4.14
Conceptual illustration of the correlation method [344].

ple source connected type multiplier [42] is employed, while in Ref. [166] a Gilbert cell [126] is employed to subtract the background light.

In this method, three-phase reference is preferable to obtain sufficient modulation information on the amplitude and the phase. Figure 4.15 shows the pixel circuits to implement the three-phase reference. The source connected circuits with three reference inputs are suitable for this purpose [344]. This gives amplitude modulation (AM)–phase modulation (PM) demodulation.

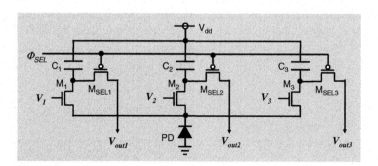

FIGURE 4.15
Pixel circuit for the correlation method. Illustrated after Ref. [344].

In the circuit, the drain current I_i in each M_i is expressed as

$$I_i = I_o exp \left(\frac{e}{mk_B T} V_i \right). \tag{4.9}$$

The notations used in the equation is the same as that in Appendix E. Each drain current of M_i, I_i is then given by

$$I_i = I \frac{\exp\left(-\frac{eV_i}{mk_BT}\right)}{\exp\left(-\frac{eV_1}{mk_BT}\right) + \exp\left(-\frac{eV_2}{mk_BT}\right) + \exp\left(-\frac{eV_3}{mk_BT}\right)}. \tag{4.10}$$

The correlation equations are thus obtained as

$$I_i - I/3 = -\frac{e}{3mk_BT}I(V_i - \bar{V}), \tag{4.11}$$

where $\bar{V}$ is the average of V_i, $i = 1, 2, 3$.

This method has been applied to a 3D range finder [340, 350] and to spectral matching [354].

4.5.3 Method of two accumulation regions

The method of using two accumulation regions in a pixel is essentially based on the same concept, but with a simpler implementation, as shown in Fig. 4.16. In this

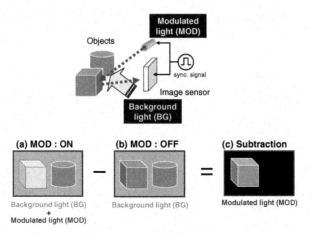

FIGURE 4.16

Concept of the two accumulation regions method.

method, the modulated signal is a pulse or ON–OFF signal. When the modulated signal is ON, the signal accumulated in one of the accumulation regions. Correlation is achieved by this operation. When the modulated signal is OFF, the signal accumulates in the other region, so that the background signal can be removed by subtracting the two signals.

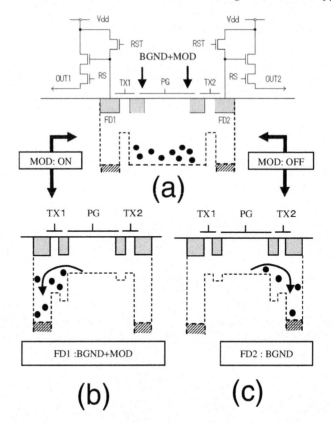

FIGURE 4.17

Pixel structure of a demodulated CMOS image sensor. The accumulated charges in the PG (a) are transferred to FD1 when the modulation light is ON (b) and transferred into FD2 when the modulation light is OFF (c).

Figure 4.17 shows the pixel structure with two accumulation regions [348]. Figure 4.18 shows the pixel layout using 0.6-μm 2-poly 3-metal standard CMOS technology. The circuit consists of a pair of readout circuits like a conventional photogate (PG) type APS, that is, two transfer gates (TX1 and TX2) and two floating diffusions (FD1 and FD2) are implemented. One PG is used as a photodetector instead of a photodiode, and is connected with both FD1 and FD2 through TX1 and TX2, respectively. The reset transistor (RST) is common to the two readout circuits. The two outputs OUT1 and OUT2 are subtracted from each other, and thus only a modulated signal is obtained. A similar structure with two accumulation regions is reported in Ref. [345, 346, 353].

The timing diagram of this sensor is shown in Fig. 4.19. First, the reset operation is achieved by turning the RST on when the modulated light is OFF. When the modulated light is turned on, the PG is biased to accumulate photocarriers. Then the

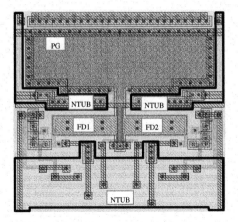

FIGURE 4.18
Pixel layout of the demodulated image sensor. The pixel size is $42 \times 42 \mu m^2$ [349].

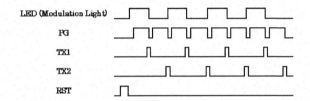

FIGURE 4.19
Timing chart of the demodulation image sensor [348]. PG: photogate, TX: transfer gate, RST: reset.

modulated light is turned off and the PG is turned off to transfer the accumulated charges to FD1 by opening TX1. This is the ON-state of the modulated light; in this state both modulated and static light components are stored in FD1. Next, the PG is biased again and starts to accumulate charges in the OFF-state of the modulated light. At the end of the OFF period of the modulated light, the accumulated charges are transferred to FD2. Thus only the static light component is stored in FD2. By repeating this process, the charges in the ON- and OFF-states accumulate in FD1 and FD2, respectively. According to the amount of accumulated charge, the voltages in FD1 and FD2 decrease in a stepwise manner. By measuring the voltage drops of FD1 and FD2 at a certain time and subtracting them from each other, the modulated signal component can be extracted.

Figure 4.20 shows experimental results obtained using the sensor [343]. One of two objects (a cat and a dog) is illuminated by modulated light and the demodulated

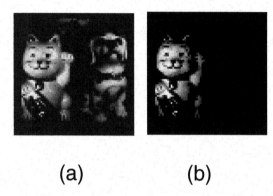

(a)   **(b)**

FIGURE 4.20
(a) Normal and (b) demodulated images [343, 349].

image only shows the cat, which is illuminated by the modulated light. Figure 4.21 shows further experimental results. In this case, a modulated LED is attached in the neck of the object (a dog), which moves around. The demodulated images show only the modulated LED and thus give a trace of the object. This means that a demodulation sensor can be applied to target tracing. Another application for a camera system suppresing saturaion is presented in Ref. [355].

This method achieves an image with little influence from the background light condition. However, the dynamic range is still limited by the capacity of the accumulation regions. In Ref. [350, 351], the background signal is subtracted in every modulation cycle so that the dynamic range is expanded. Although the adding circuits consume pixel area, this technique is effective for demodulation CMOS image sensors.

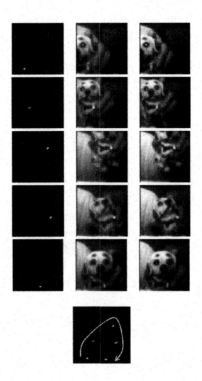

FIGURE 4.21

Demonstration of marker tracking. The images are placed in order of time from top to bottom. Left column: modulated light pattern extracted by the sensor. Middle column: output from the modulated light and background light. Right column: output from only the background light. The bottom figure shows the moving trace of the marker. For convenience, the moving direction and the track of the LED are superimposed [343].

4.6   Three-dimensional range finder

A range finder is an important application for factory automation (FA), ITS, robot vision, gesture recognition, etc. By using smart CMOS image sensors, three-dimensional (3D) range finding or image acquisition associated with distances can be realized. Several approaches suitable for CMOS image sensors have been investigated. Their principles are based on time-of-flight (TOF), triangulation, and other methods, summarized in Table 4.2. Figure 4.23 shows images taken with a 3D range finder [356]. The distance to the object is shown on the image.

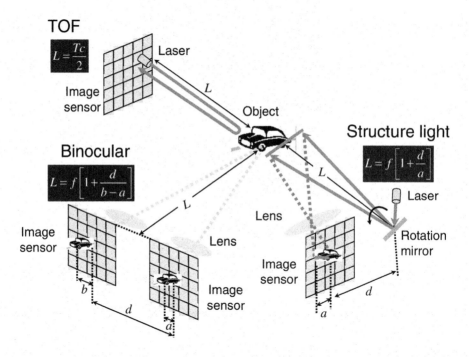

FIGURE 4.22
Concept of three methods for 3D range finders: TOF, binocular, and light-section.

4.6.1   Time of flight

TOF is a method to measure the round-trip time of flight and has been used in light detection and ranging (LIDAR) for many years [377]. The distance to the object L is

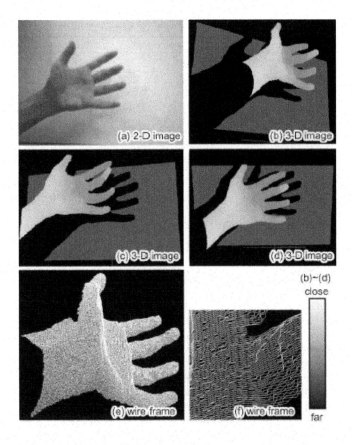

FIGURE 4.23
Examples of images taken by a 3D range finding image sensor developed by Y. Oike et al. [356]. Courtesy of Dr. Y. Oike.

expressed as

$$L = \frac{Tc}{2}, \tag{4.12}$$

where T is a round-trip time (TOF = $T/2$) and c is the speed of light. The most notable feature of TOF is its simple system; it requires only a TOF sensor and a light source. TOF sensors are classified by direct and indirect TOF.

4.6.1.1 Direct TOF

A direct TOF sensor measures the round-trip time of light in each pixel directly. Consequently, it requires a high speed photodetector and high precision timing circuits. For example, for $L = 3$ m, $T = 10^{-8}$ sec= 10 psec. To obtain mm accuracy, an

TABLE 4.2

Smart CMOS sensors for 3D range finders

Category	Implementation method	Affiliation and Refs.
Direct TOF	APD array	ETH [97, 98], MIT [357]
Indirect TOF	Pulse	ITC-irst [358, 359], Fraunhofer [360, 361], Sizuoka U. [362]
	Sinusoidal	PMD [340, 341, 363, 364], C-SEM [353, 365], Canesta [366], JPL [367]
Triangulation	Binocular	Shizuoka U. [368], Tokyo S-ci. U. [369] Johns Hopkins U. [370, 371]
	Structured light	Carnegie Mellon U. [87, 372] U. Tokyo [350, 356, 373, 374], SONY [157], [375], Osaka EC. U. [342]
Others	Light intensity (Depth of intensity)	Toshiba [376]

averaging operation is necessary. The advantage of direct TOF is its wide range for measuring distance, from meters to kilometers.

As high speed photodetectors in standard CMOS technology, APDs are used in the Geiger mode for direct TOF sensors [97, 98, 357], as discussed in Sec. 2.3.4. A TOF sensor with 32×32 pixels, each of which is integrated with the circuits in Fig. 4.6 in Sec. 4.2.4 with an area of $58 \times 58 \ \mu m^2$, has been fabricated in 0.8-μm standard CMOS technology with a high voltage option. The anode of the APD is biased at a high voltage of -25.5 V. The jitter of the pixel is 115 ps, so that to obtain mm accuracy an averaging operation is necessary. A standard deviation of 1.8 mm is obtained with multiple depth measurements with 10^4 at a distance of around 3 m.

4.6.1.2 Indirect TOF

To alleviate the requirements for direct TOF, indirect TOF has been developed [340, 341, 345, 353, 358–361, 363–367]. In indirect TOF, the round-trip time is not measured directly, but two modulated light signals are used. An indirect TOF sensor generally has two accumulation regions in each pixel to demodulate the signal, as mentioned in Sec. 4.5. The timing diagram of indirect TOF is illustrated in Fig. 4.24. In this figure, two examples are shown.

When the modulation signal is a pulse or an on/off signal, two pulses with a delay time t_d between them are emitted with a repetition rate of the order of MHz. Figure 4.25 illustrates the operation principle [358, 359]. In this method, the TOF signal is obtained as follows. Two accumulation signals V_1 and V_2 correspond to the two

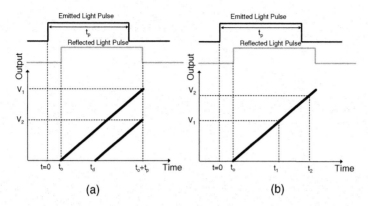

FIGURE 4.24

Timing diagram of indirect TOF with two different pulses. (a) The second pulse has a delay of t_d against the first pulse. (b) Two pulses with different durations [359].

pulses with the same width t_p, as shown in Fig. 4.24(a). From the delay time t_d and the two times of TOF, the distance L is computed as

$$t_d - 2 \times TOF = t_p \frac{V_1 - V_2}{V_1} \tag{4.13}$$

$$TOF = \frac{L}{c} \tag{4.14}$$

$$\therefore L = \frac{c}{2} \left[t_p \left(\frac{V_2}{V_1} - 1 \right) + t_d \right]. \tag{4.15}$$

In Ref. [359], an ambient subtraction period is inserted in the timing and thus a good ambient light rejection of 40 klux is achieved. A 50×30-pixel sensor with a pixel size of $81.9 \times 81.7\ \mu m^2$ has been fabricated in standard 0.35-μm CMOS technology and a precision of 4% is obtained in the 2—8 m range.

Another indirect TOF with pulse modulation uses two pulses with different timing, T_1 and T_2 [360–362]. The procedure to estimate L is as follows. V_1 and V_2 are output signal voltages for the shutter time t_1 and t_2, respectively. As shown in Fig 4.24(b), from these four parameters, the intersect point $t_o = TOF$ can be interpolated. Consequently:

$$L = \frac{1}{2} c \left(\frac{V_2 t_1 - V_1 t_2}{V_2 - V_1} \right). \tag{4.16}$$

Next, a sinusoidal emission light is introduced instead of the pulse for indirect TOF [340, 341, 353, 363–367].

The TOF is obtained by sampling four points, each of which is shifted by $\pi/2$, as shown in Fig. 4.26, and calculating the phase shift value ϕ [363]. The four sampled values A_1–A_4 are expressed by the signal phase shift ϕ, the amplitude a, and the

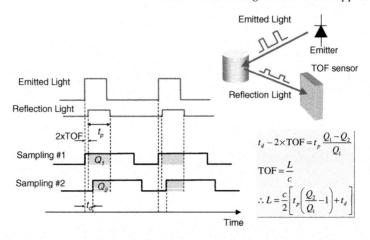

FIGURE 4.25
Operation principle of indirect TOF with two emission pulses with a delay.

offset b as

$$A_1 = a\sin\phi + b, \tag{4.17}$$
$$A_2 = a\sin(\phi + \pi/2) + b, \tag{4.18}$$
$$A_3 = a\sin(\phi + \pi) + b, \tag{4.19}$$
$$A_4 = a\sin(\phi + 3\pi/2) + b. \tag{4.20}$$

From the above equations, ϕ, a, and b are solved as

$$\phi = \arctan\left(\frac{A_1 - A_3}{A_2 - A_4}\right). \tag{4.21}$$

$$a = \frac{\sqrt{(A_1 - A_3)^2 + (A_2 - A_4)^2}}{2}. \tag{4.22}$$

$$b = \frac{A_1 + A_2 + A_3 + A_4}{4}. \tag{4.23}$$

Finally, by using ϕ, the distance L can be calculated as

$$L = \frac{c\phi}{4\pi f_{mod}}, \tag{4.24}$$

where f_{mod} is the repetition frequency of the modulation light.

To realize indirect TOF in a CMOS sensor, several types of pixels have been developed with two accumulation regions in a pixel. They are classified into two techniques: one is to place two FDs on either side of the photodetector [340,341,345,353,

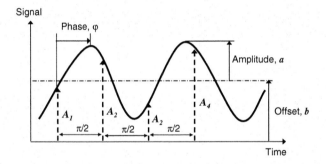

FIGURE 4.26
Operation principle of indirect TOF with sinusoidal emission light. Illustrated after Ref. [363].

362, 363, 365, 366], the other is to use a voltage amplifier to store signals in capacitors [358–361]. In other words, the first technique is to change the photodetection device, while the second is to use conventional CMOS circuitry. The photomixing device (PMD) is a commercialized device that has a PG with two accumulation regions on either side [363]. The maximum available pixel number is 160×120. Another commercialized sensor for indirect TOF has a 64×64 pixel array with a high-speed clock generator and an ADC [366]. The sensors in Refs. [366] and [362] are fabricated in a standard CMOS process or a slightly modified process. In particular, the sensor reported in Ref. [362] has a QVGA array, which has a wide range of applications such as ITS and gesture recognition.

4.6.2 Triangulation

Triangulation is a method to measure the distance to the field of view (FOV) by a triangular geometrical arrangement. This method can be divided into two classes: passive and active. The passive method is also called the binocular or stereo-vision method. In this method two sensors are Used. The active method is called the structured light method. In this method a patterned light source is used to illuminate the FOV.

4.6.2.1 Binocular method

The passive method has the advantage that it requires no light sources and only two sensors are needed. Reports of several such sensors have been published [368–371]. The two sensors are integrated into two sets of imaging area to execute stereo vision. However, this means that the method must include a technique to identify the same FOV in the two sensors, which is a very complicated problem for typical scenes. For the sensor reported in Ref. [368], the FOV is restricted to a known object and three-dimensional information is used to improve the recognition rate of the object. The

sensor has two imaging areas and a set of readout circuitry including ADCs. The
sensor in Ref. [371] integrates the current mode disparity computation circuitry.

4.6.2.2 Structured light method

Considerable research on the active method has been published [87, 157, 342, 350,
356, 372–374]. In the active method, a structured light source is required, which
is usually a stripe-shaped light source, The light source is scanned over the FOV.
To distinguish the projected light pattern from the ambient light, a high power light
source with a scanning system is required. In Ref. [372], the first integrated sensor
with 5×5 pixels was reported for 2-μm CMOS technology.

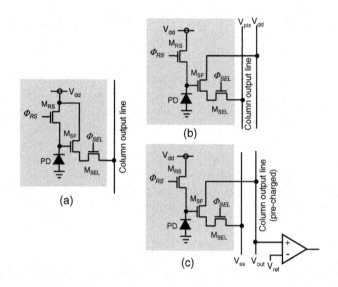

FIGURE 4.27
(a) Conventional 3T-APS structure, modified for (b) normal imaging mode, (c) PWM
mode. Illustrated after Ref. [373].

In the structure light method, finding a pixel at the maximum value is essential. In
Ref. [87], a winner-take all (WTA) circuit is used to find the maximum. The WTA
circuit is mentioned in Sec. 3.3.1. In Refs. [356, 373], a conventional 3T-APS is
used in two modes, the normal image output mode and the PWM mode, as shown in
Fig. 4.27. PWM is discussed in Sec. 3.4. PWM can be used as a 1-bit ADC in this
case. In the sensor, PWM-ADCs are located in the column and thus column-parallel
ADC is achieved. In Fig. 4.27(b), the pixel acts as a conventional 3T-APS, while in
Fig. 4.27(c), the output line is precharged and the output from a pixel is compared
with the reference voltage V_{ref} in the column amplifier. The output from the pixel

decreases in proportion to the input light intensity so that PWM is achieved. Using a 3T-APS structure, the sensor achieves a large array format of VGA with a good accuracy of 0.26 mm at a distance of 1.2 m. Refs. [157] and [375] report on the implementation of analog current copiers [378] and comparators to determine the maximum peak quickly. In Ref. [157], analog operation circuits are integrated in a pixel, while in Ref. [375], four sets of analog frame memory are integrated on a chip to reduce the pixel size and a color QVGA array is realized with a 4T-APS pixel architecture.

In the structured light method, it is also important to suppress the ambient light against the structured light, which is incident on the FOV. Combined with a wide dynamic range using a log sensor and modulation technique, Ref. [350] reports a good signal-to-background ratio (SBR) of 36 dB with a DR of 96 dB*.

4.6.3 Depth key

The depth key method refers to "depth of focus," "depth of gray-scale," and so on [379]. It is a method to measure the distance for the depth of some physical parameter. The method is often used in a camera system, as is discussed in Ref. [380]. A depth of focus sensor can calculate the following Laplacian based value d for whole pixels on a chip from several pictures with different focuses.

$$d = |l_{F(x+1,\,y)} + l_{F(x-1,\,y)} - 2l_{F(x,\,y)}| + |l_{F(x,\,y+1)} + l_{F(x,\,y-1)} - 2l_{F(x,\,y)}| \quad (4.25)$$

In simulation results, a system can output a depth map (64×64 pixels and 50 depth steps) at 30 fps, when images are taken at 1500 fps using $0.6\text{-}\mu$m standard CMOS technology [380].

The depth of gray-scale method has also been reported [376]. A depth of gray-scale sensor can measure the distance to an object by the intensity reflected from the object, so that the evaluated distance is affected by the reflectance of the object. This method is easy to implement.

4.7 Target tracking

Target tracking or target tracing is a technique to track specific object(s). It is an important application for smart CMOS image sensors. It requires real-time signal processing; for example, robot vision requires fast and compact processing as well as low power consumption. Many smart CMOS smart image sensors have been reported so far and in this section, we classify several types of smart CMOS image sensors that can realize target tracking and discuss the operation principle.

---

*In Ref. [350], the definition of SBR and DR are different by a half factor from the conventional definition. In this book, the conventional definition is adopted.

To achieve target tracking, it is necessary to extract the object to be tracked. For this purpose, it is important to find the centroid of the targeted object. There are several reports on estimating the centroid of an object in a scene. In acquiring the centroid, frame difference operation is effective because moving objects can be extracted with this operation. To implement the frame difference operation, a frame buffer is usually required, while in some smart sensors frame memory is employed in a pixel [328, 381] so that no frame memory outside the chip is needed. In Ref. [328], the difference of the pixel values between two frames was monitored on the chip to find moving objects. Ref. [382] reports employing an in-pixel capacitor and comparator to detect moving objects. To detect movement, a resistive network architecture is also effective [42, 45], discussed in Sec. 3.3.3.

Target tracking sensors are mainly classified into analog processing and digital processing. Table 4.3 summarizes the methods with some typical examples. Each method requires some pre-processing of input images, such as edge detection, binarization, etc. Analog processing allows the use of maximum detection ("MaxDet." in Table 4.3), projection, and resistive network ("RN" in Table 4.3).

Another method for target tracking is the modulation technique, which is discussed in Sec. 4.5. In this case, although a modulation light source is required, the tracking is relatively easy because it easy to distinguish the region-of-interest (ROI) from other objects in this method.

4.7.1   Maximum detection for target tracking

Maximum detection (MaxDet.) is a method to detect the maximum pixel value in the image containing the target to be traced. In Ref. [383], WTA circuits are employed in each pixel to detect the maximum pixel value in the image. WTA is discussed in Sec. 3.3.1. To specify the x–y position of the maximum value, two sets of one-dimensional resistive networks placed in rows and columns are used. A parasitic phototransistor (PTr) is integrated with the WTA circuits in a pixel with a size of 62μm $\times$ 62μm. The chip has 24×24 pixels and demonstrates a processing speed of 7000 pixel/sec. Another method of processing for MaxDet. is to use a one-dimensional resistive network in row and column directions combined with a comparator to report the position of the maximum in each direction [384].

4.7.2   Projection for target tracking

Several reports using projection have been published [173, 385–387]. Projection is easy to implement because only summations along the rows and columns are required. Projection along each row and column direction is a first-order image moment. Preprocessing, such as edge detection [385] and binarization [386], is effective to obtain the centroid clearly. In Ref. [385], the image area is divided into two areas with different pixel densities: the fovea and peripheral areas. This structure is a mimic of the distribution of photoreceptors in the human retina. The fovea area with dense pixel density performs edge detection and motion detection, while the peripheral area with sparse pixel density performs edge detection and projection.

The photodetection structure is similar to the log sensor discussed in Sec. 3.2.2, that is, a PTr with a subthreshold transistor that produces a logarithmic response. The edge detection is achieved by a resistive network.

A smart CMOS image sensor that simultaneously produces a normal image and projections along the row and column directions has been reported [387]. In this sensor, a 20-μm square pixel has three PDs, one of which is for imaging and the others are used for projections. An APS is used for imaging and a passive pixel sensor (PPS) is used for projection, which processes the sum of the charges from pixels along one direction. This sensor also has a global shutter and random access functions and thus is effective for high-speed image processing. The sensor has 512×512 pixels and was fabricated in 0.6-μm standard CMOS technology.

4.7.3 Resistive network and other analog processing for target tracking

Another analog processing method is the resistive network. In [45], a two-layered resistive network architecture is implemented, as shown in Fig. 3.7, and a spatio-temporal difference is realized using the frame difference method. Target tracking is achieved by using median filtering outside the chip. By introducing an APS and canceling circuits, this sensor has achieved a good SNR with low FPN in a resistive network architecture. The architecture of this sensor is described in Sec. 3.3.3.

Another analog processing method using fully parallel processing based on a cellular neural network (CNN) has been reported [388]. Analog processing and asynchronizing digital circuits are integrated in a pixel with an area of 85 μm square using 0.18-μm standard CMOS technology. The number of pixels is 64×64. The chip achieves a processing time of 500 μsec and $\sim$725 MIPS/mW. Using a fully analog architecture, a very low power consumption of 243 μW is achieved under a 1.8-V power supply.

4.7.4 Digital processing for target tracking

Finally, we discuss digital processing chips. In these chips, one-bit digital processing unit or bit-serial processing is integrated in a pixel with an area size of 80 μm square using 0.5-μm standard CMOS technology. The block diagram of a chip is shown in Fig. 4.28. The pixel circuits have been shown earlier in Fig. 3.18. As fully digital processing is implemented, the chip is fully programmable and achieves very fast processing of around 1 msec. Such digital processing chips require an ADC in each pixel; in this chip a PWM-like ADC is introduced by using a simple inverter, as described in Sec. 3.4.1.

By using the chip, fast target tracking has been demonstrated for a micro-visual feedback system combined with depth-of-focus applied to the observation of a moving biological specimen, as shown in Fig. 4.29 [389]. It is difficult to track a moving object by observing with an optical microscope, and thus it would be useful to be able to automatically track a small object using an optical microscope system. Experiments reported in Ref. [389] show excellent tracking results using a digital vision

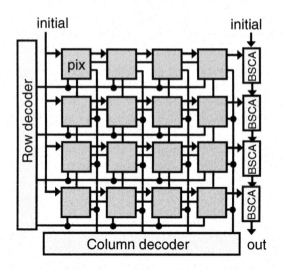

FIGURE 4.28
Chip block diagram of a fully digital smart CMOS image sensor [61]. BSCA: bit-serial cumulative accumulator.

chip with a 1-kHz operation speed combining depth-of-focus algorithms to achieve three-dimensional tracking.

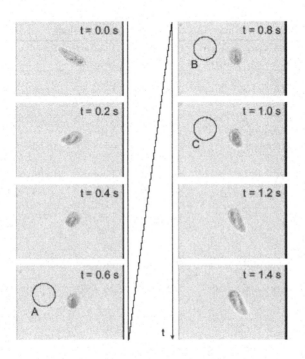

FIGURE 4.29

Tracking results for a moving biological specimen. The images were taken by a CCD camera attached to an optical microscope. The tracking was achieved by a digital vision chip [389]. Courtesy of Prof. M. Ishikawa at the Univ. of Tokyo.

TABLE 4.3
Smart CMOS sensors for target tracking

Method	Preprocess	Tech. (μm)	Pixel #	Pixel size (μm)	Speed	Power consum.	Ref.
MaxDet.	WTA & 1D-RN	2	24 × 24	62 × 62	7 kpix/s		[383]
MaxDet.	1D-RN & comparator	0.6	11 × 11	190 × 210	50 μs (sim.)		[384]
Projection	Current sum	2	256 × 256	35 × 26	10 ms		[173]
Projection	Edge detect	2	Fovea 9 × 9, periphery 19 × 17	Fovea 150 × 150, periphery 300 × 300	10 lpix/s	15 mW	[385]
Projection	Binarization	0.18	80 × 80	12 × 12	1000 fps	30 mW	[386]
Projection	Charge sum	0.6	512 × 512	20 × 20	1620 fps	75 mW	[387]
2D-RN		0.25	100 × 100 (hex.)	87 × 75.34	~7 μs	36.3 mW	[45]
CNN		0.18	48 × 48	85 × 85	500 μs, ~725 MIPS/mW	243 μW	[388]
Digital process.	ADC	0.5	64 × 64	80 × 80	1 kHz	112 mW	[61]

4.8 Dedicated arrangement of pixel and optics

This section describes smart CMOS image sensors that have dedicated pixel arrangements and related optics. A conventional CMOS image sensor uses a lens to focus an image onto the image plane of the sensor, where pixels are placed in an orthogonal configuration. In some visual systems, however, non-orthogonal pixel placement is used. A good example is our vision system where the distribution of photoreceptors is not uniform; around the center or fovea, they are densely placed, while in the periphery they are sparsely distributed [187]. This configuration is preferable as it is able to detect an object quickly with a wide angle view, then once the object is located, it can be imaged more precisely by the fovea by making the eye move to face the object. Another example is an insect with compound eyes [390, 391]. A special placement of pixels is sometimes combined with special optics, such as in the compound eyes of insects. Pixel placement is more flexible in a CMOS image sensor than in a CCD sensor because the alignment of CCDs is critical for charge transfer efficiency; for example, a curved placement of a CCD may degrade the charge transfer efficiency.

This section first describes special pixel arrangements for smart CMOS image sensor. Then, some smart CMOS sensors with dedicated optics are introduced.

4.8.1 Non-orthogonal arrangement

4.8.1.1 Foveated sensor

A foveated sensor is an image sensor inspired by the human retina. In a human retina, photoreceptors are arranged so that their density around the center or fovea is larger than in the periphery [392].

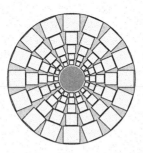

FIGURE 4.30
Pixel arrangement of a foveated image sensor. The pixel size increases by square root, that is the pixel pitch logarithmically decreases. The circle area of the center shows the control circuits.

A foveated sensor has a pixel arrangement like the human retina; pixels are arranged logarithmically decreasing in pixel density along the radial direction, as shown in Fig. 4.30. This is called the log-polar coordination [393–396]. In the log-polar coordination, the Cartesian coordinates are converted into log-polar coordinates by

$$r = \sqrt{x^2 + y^2}$$
$$\theta = \arctan\left(\frac{y}{x}\right).$$

(4.26)

The foveated or log-polar image sensor is useful for some image processing, but the layout of the sensor has some difficulties. First, it is difficult to place pixels around the center due to the high pixel density and thus no image can be taken around the center. Second, the placement of the two scanners for the radial and circular directions is problematic. To alleviate these problem, R. Etienne-Cummings et al. have developed a foveated image sensor employing only two types of central and peripheral areas, placed in an orthogonal arrangement or a conventional arrangement [385]. The sensor is applied to target tracking, discussed in Section 4.7. The placement problem of the scanners is the same as for hyper omni vision, which is discussed in the next section.

4.8.1.2 Hyper omni vision

Hyper omni vision (HOVI) is an imaging system that can capture a surrounding image in all directions by using a hyperbolic mirror and a conventional CCD camera [379, 397]. The system is suitable for surveillance. The output image is projected by a mirror and thus is distorted. Usually, the distorted image is transformed to a image rearranged with Cartesian coordinates and is then displayed. Such an off-camera transformation operation restricts available applications. A CMOS image sensor is versatile with regard to pixel placement. Thus, pixels can be configured so as to adapt for a distorted image directly reflected by a hyperbolic mirror. This realizes an instantaneous image output without any software transformation procedure, which open up various applications. In this section, the structure of a smart CMOS image sensor for HOVI and the characteristics of the sensor are described [398].

A conventional HOVI system consists of a hyperbolic mirror, a lens, and a CCD camera. Images taken by HOVI are distorted due to the hyperbolic mirror. To obtain a recognizable image, a transformation procedure is required. Usually this is done by software in a computer. Figure 4.31 illustrates the imaging principle of HOVI.

An object located at $P(X,Y,Z)$ is projected to a point $p(x,y)$ in the two-dimensional image plane by the hyperbolic mirror. The coordinates of $p(x,y)$ are expressed as follows.

$$x = \frac{Xf\left(b^2 - c^2\right)}{(b^2 + c^2)Z - 2bd\sqrt{X^2 + Y^2 + Z^2}},$$

(4.27)

$$x = \frac{Yf\left(b^2 - c^2\right)}{(b^2 + c^2)Z - 2bd\sqrt{X^2 + Y^2 + Z^2}}.$$

(4.28)

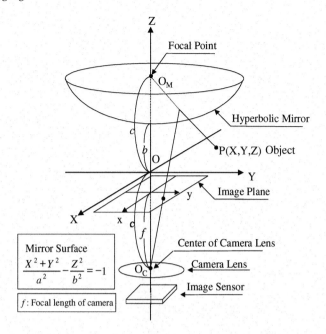

FIGURE 4.31
Configuration of a HOVI system.

Here, b and c are parameters of the hyperbolic mirror and f is the focal length of the camera.

Figure 4.33(a) illustrates image acquisition for conventional HOVI with a CCD camera. The output image is distorted. A smart CMOS image sensor is designed to arrange pixels in a radial pattern according to Eqs. 4.27 and 4.28 [398]. The sensor is fabricated in 0.6-μm 2-poly 3-metal standard CMOS technology. The specification of the fabricated chip is described in Table 4.4.

TABLE 4.4
Specifications of a smart CMOS image sensor for HOVI

Technology	0.6-μm 2-poly 3-metal standard CMOS
Chip size	8.9 μm sq.
Pixel number	32 $\times$ 32
PD structure	N-diff./P-sub.
Power supply	5 V
PD size	18 μm sq. from 1st to 8th pixel along a radial direction
	20 μm sq. from 9th to 16th
	30 μm sq. from 17th to 24th
	40 μm sq. from 25th to 32nd

A 3T-APS is used for the pixel circuits. A feature of the chip is that the pitch of the pixel becomes smaller from the outside edge to the center. Thus, four types of pixels with different area sizes are employed as, described in Table 4.4. For the radial configuration, vertical and horizontal scanners are placed along the radial and circular directions, respectively. A microphotograph of the fabricated chip is shown in Fig. 4.32. In the figure, a close-up view around the inner most pixels is also shown.

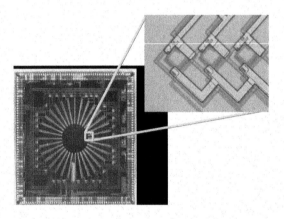

FIGURE 4.32
Microphotograph of a smart CMOS image sensor for a HOVI system. A close-up microphotograph is also shown.

Figure 4.33 shows experimental results for this sensor. The input pattern is an image taken by a conventional HOVI camera system with a CCD camera. This result clearly shows that the output from the fabricated image sensor restores the original input image taken by the HOVI camera system.

4.8.2 Dedicated optics

4.8.2.1 Compound eye

A compound eye is a biological visual systems in arthropods including insects and crustaceans. There are a number of independent tiny optical systems with small fields-of-view (FOV) as shown in Fig. 4.34. The images taken by each of the independent tiny eyes, called *ommatidium*, are composited in the brain to reproduce a whole image.

The advantages of a compound eye are its wide FOV with a compact volume and a short working distance, which can realize an ultra thin camera system. Also, only a simple imaging optics is required for each ommatidium, because only a small FOV is required for an ommatidium. The disadvantage is relatively poor resolution.

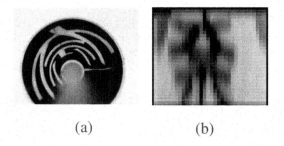

(a)　　　　　　　　(b)

FIGURE 4.33
Input of a Japanese character taken by a conventional HOVI system and its output images for the proposed sensor.

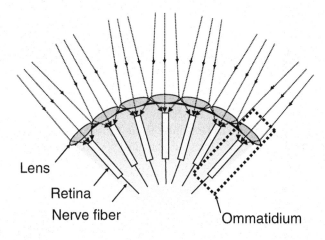

FIGURE 4.34
Concept of an apposition compound eye system. The system consists of a number of ommatidium, which are composed of a lens, retina, and nerve fiber. It is noted that another kind of compound eye is the neural superposition eye [390].

Artificial compound eyes have been developed by many institutes [399–404]. In the next sections, two examples of smart image sensor systems using compound eye architecture are introduced, DragonflEye by R. Hornsey et al. of York University and TOMBO by J. Tanida et al. of Osaka University.

DragonflEye "DragonflEye" is a compound eye image sensor where up to 20 eyelets or ommatidiums are implemented and focused on the sensor imaging plane with approximately 150-pixel resolution in each eyelet [404]. It mimics an eye system of a dragonfly and is intended for applications in high-speed object tracking and depth perception. Figure 4.35 shows a prototype system. The optical system realizes a wide angle view. By using a smart CMOS image sensor with a random access function, high-speed accessing of each sub-imaging region is achieved.

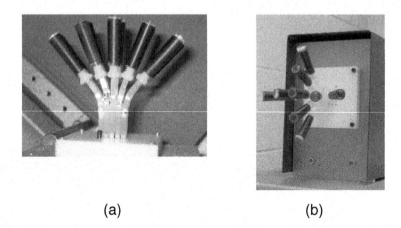

(a)   (b)

FIGURE 4.35
Photographs of the prototype of DragonflEye. (a) Close up of eyelet bundle, (b) total system [404]. Each eyelet has a lens and a bundle fiber. Courtesy of Prof. R. Hornsey at York Univ.

TOMBO   The TOMBO* system, an acronym for thin observation module by bound optics, is another compound eye system [405, 406]. Figure 4.36 shows the concept of the TOMBO system. The heart of the TOMBO system is the introduction of a number of optical imaging systems, each of which consists of several micro-lenses, called imaging optical units. Each imaging optical unit captures a small but full image with a different imaging angle. Consequently, a number of small images with different imaging angles are obtained. A whole image can be reconstructed from the compound images from the imaging optical units. A digital post processing algorithm enhances the composite image quality.

To realize a compound eye system, a crucial issue is the structure of the ommatidium with micro-optics technology. In the TOMBO system, the signal separator shown in Fig. 4.36 resolves the issue. A CMOS image sensor dedicated for the TOMBO system has been developed [118, 407].

The TOMBO system can also be used as a wide angle camera system as well as a thin or compact camera system. In this system, 3 × 3 units are employed with a single lens. Figure 4.37(a) shows the system structure and 3 × 3 unit images taken

---

*TOMBO is the Japanese word for dragonfly. A dragonfly has a compound eye system.

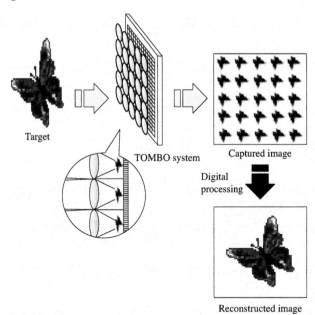

FIGURE 4.36
Concept of TOMBO. Courtesy of Prof. J. Tanida at Osaka Univ.

by this camera. By attaching two prisms, a wide angle view of 150 degrees can be obtained, as shown in Fig. 4.37(b). Each optical imaging unit has a small scanning area and hence a rolling shutter, as discussed in Sec. 2.11, does not cause serious distortion of the total image. By taking images with each unit at a different time, a moving object can be detected, as shown in Fig. 4.37(c).

4.8.2.2 Polarimetric imaging

Polarization is a characteristic of light [391]. Polarimetric imaging makes use of polarization and is used in some applications to detect objects more clearly. Humans cannot sense polarization, though some animals such as bees can sense it. Several types of polarimetric image sensors has been developed by placing birefringent materials on the sensor surface [408]. Such types of smart CMOS image sensor may be useful for chemical applications where polarization is frequently used to identify chemical substances.

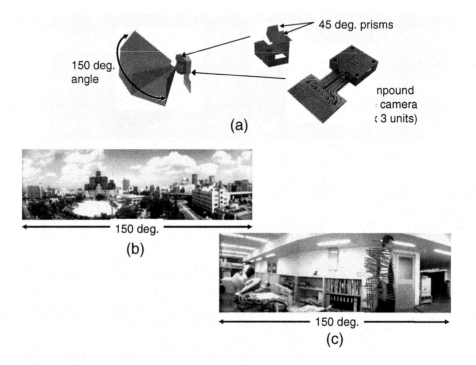

(a)

(b)

(c)

FIGURE 4.37
Wide angle view camera system using the TOMBO system. (a) The camera system consists of a TOMBO camera with 3 × 3 units and two sets of prism with an attachment holder. (b) Obtained wide angle view. (c) Detection of moving objects. Courtesy of Mr. Masaki and Mr. Toyoda of Funai Electric Co. Ltd.

5

Applications

5.1 Introduction

In this chapter, several applications for smart CMOS image sensors are introduced.

The first applications considered are those for communications and information, including human interfaces. Introducing imaging functions in these fields can improve not only performance, such as communication speed, but also convenience, such as in visually aided controls.

The second category of applications considered is in the biotechnology fields. In these fields, CCDs combined with optical microscopes are widely used. If smart CMOS image sensors are applied, the imaging systems could be made compact or have integrated functions, leading to improvements in performance.

Finally, medical applications are introduced. Medical applications require very compact imaging systems, because they may be introduced into the human body through swallowing or being implanted. Smart CMOS image sensors are suitable for these applications because of their small volume, low power consumption, integrated functions, etc.

5.2　Information and communication applications

Since blue LEDs and white LEDs have emerged, light sources have drastically changed; for example, some room lighting, lamps in automobiles, and large outdoor displays use LEDs. The introduction of LEDs with their fast modulation has produced a new application, free-space optical communication. Combined with image sensors, free-space optical communication can be extended to human interfaces because images are visually intuitive for humans.

In this section, we introduce information and communication applications of smart CMOS image sensors. First, an optical identification (ID) tag is introduced. Next, optical wireless communication is described.

5.2.1   Optical ID tag

The recent spread of large outdoor LED displays, white LED lighting, etc. has prompted research and development into applying LEDs to transceivers in spatially optical communication systems. Visible light communication is one such system [409]. This application is described in Sec. 5.2.2. An alternative application is

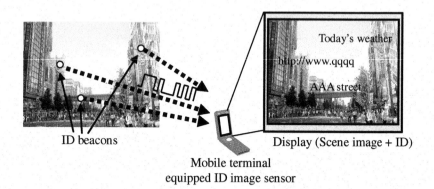

ID beacons   Display (Scene image + ID)
Mobile terminal
equipped ID image sensor

FIGURE 5.1
Concept of optical ID tag. LEDs on a large LED display are used as optical ID tags. An image that a user takes is superimposed ID data.

an optical ID tag, which uses an LED as a high-speed modulator to send ID signals. Commercially available optical ID tags have been used for motion capture [410,411]. Recently, applying optical ID tags to augmented reality technologies has been proposed and developed such as the following systems: "Phicon" [412], "Navi cam" [413], "ID cam" [414,415], "A*i*mulet" [416,417], and "OptNavi" [418–420].

Phicon, A*i*mulet   These systems have been developed for sending and receiving data to a user. A user has a terminal to keep track of his or her location and to send/receive data from a server station. The Phicon system was proposed by D.J. Moore at GITech and R. Want et al. at Xerox PARC [412]. It utilizes an IR LED as an optical beacon to relay a user's location and send data. A monochrome CCD is used to locate optical beacons and decode the transmitted data from IR transceivers. The bit rate is about 8 bps due to the use of a conventional CCD camera for decoding the modulated light. This suggests that a dedicated camera system is required for detecting beacons with a faster data rate.

A*i*mulet was developed by H. Itoh at AIST,* Japan. It is used as a handy communicator for a user to find information in museums and exhibitions . A few demon-

---
*Advanced Institute of Science and Technology.

stration models have been developed and tested in the large Aichi Expo exhibition in Japan.

ID cam, OptNavi systems The ID cam, proposed by Matsushita et al. at Sony, uses an active LED source as an optical beacon with a dedicated smart CMOS image sensor to decode the transmitted ID data. The smart sensor used in the ID cam is described later.

OptNavi has been proposed by the author's group at NAIST,* Japan [418] and is designed for use with mobile phones as well with the ID cam. One typical application of these systems, LEDs on a large LED display, can be used as an optical beacon, as shown in Fig. 5.1. In the figure, LED displays send their own ID data. When a user takes an image using a smart CMOS image sensor that can decode these IDs, the decoded data is superimposed on the user interface, as shown in Fig. 5.1. The user can easily get information about the contents on the displays.

An alternative application is a visual remote controller for electronic appliances connected together in a network. Interest in home networks for electronic appliances is growing and some consortiums have been established, such as DLNA®* [421], ECHONET* [422], UPnP™* Forum [423], and HAVi* [424]. Many networked appliances can be connected to each other over a home network. The OptNavi system has been proposed for use in a man–machine interface to control networked appliances visually. In this system, a mobile phone with a custom image sensor is used as an interface; many mobile phones are equipped with a large display, a digital still camera, IrDA [425], Bluetooth® [426], etc. In the OptNavi system, home network appliances, such as a DVD recorder, a TV, and a PC, are equipped with LEDs that transmit ID signals at 500 Hz. The image sensor can receive IDs with a high-speed readout of multiple region-of-interests (ROIs). The received IDs are displayed with a superimposed background image captured by the sensor, as shown in Fig. 5.2. With the OptNavi system, we can control such appliances by visually confirming them on a mobile phone display.

5.2.1.1 Smart CMOS image sensors for optical ID tags

Since a conventional image sensor captures images at a video frame rate of 30 fps, it can not receive ID signals at a kHz rate. Thus, a custom image sensor with a function for receiving ID signals is needed. Several smart CMOS image sensors dedicated for optical ID tags have been proposed and demonstrated [157, 375, 427– 431], summarized in Table 5.1. These sensors receive ID signals with a high speed frame rate. S. Yoshimura et al. at Sony and T. Sugiyama et al. from the same group have demonstrated image sensors that operate with high speed readout for all pixels

---

*Nara Institute of Science and Technology.

*Digital Living Network Alliance.

*Energy Conservation and Homecare Network.

*Universal Plug and Play.

*Home Audio Video Interoperabitily.

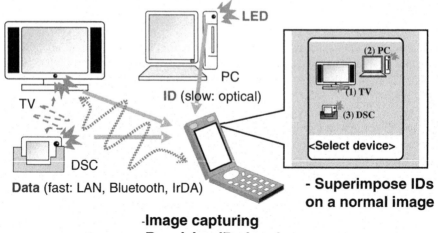

- Superimpose IDs on a normal image

-Image capturing
-Receiving ID signals

FIGURE 5.2

Concept of OptNavi. A dedicated smart CMOS image sensor can detect and decode
optical IDs from home electronic appliances. The decoded results are superimposed
on the image taken by the sensor.

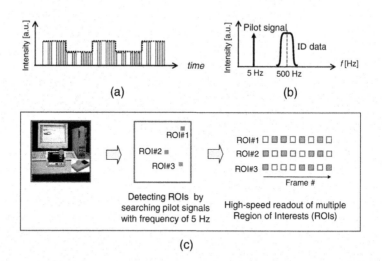

FIGURE 5.3

Method of detecting ID positions using a slow pilot signal. (a) Waveforms of pilot
and ID signals, (b) power spectrum of pilot and ID signals, (c) procedure to obtain
an ID signal using a pilot signal.

with the same pixel architecture as a conventional CMOS image sensor [157, 375]. These sensors are suitable for high resolution.

Oike et al. at Univ. Tokyo has demonstrated a sensor that receives ID signals by analog circuits in a pixel, capturing images at a conventional frame rate [428]. This sensor can receive ID signals with high accuracy.

TABLE 5.1
Specifications of smart CMOS image sensors for optical ID tags

Affiliation	Sony	Univ. Tokyo	NAIST
Reference	[375]	[427]	[431]
ID detection	High speed readout of all pixels	In-pixel ID receiver	Readout of multiple ROIs
Technology	0.35-μm CMOS	0.35-μm CMOS	0.35-μm CMOS
Pixel size	11.2 × 11.2 μm^2	26 × 26 μm^2	7.5 × 7.5 μm^2
Number of pixels	320 × 240	128 × 128	320 × 240
Frame rate for ID detection	14.2 kfps	80 kfps	1.2 kfps
Power consumption	82 mW (@3.3 kfps, 3.3 V)	682 mW (@4.2 V)	3.6 mW (@ 3.3 V, w/o ADC, TG)
Issues	Power consumption	Large pixel size	Speed

High-speed readout of multiple ROIs with low power consumption   High-speed readout of all pixels may cause large power consumption. The NAIST group has proposed an image sensor dedicated for optical ID tags to realize high-speed readout at low power consumption with a simple pixel circuit [419]. In the readout scheme, the sensor is operated for capturing normal images at a conventional video frame rate while simultaneously capturing multiple ROIs which receive ID signals at a high-speed frame rate. To locate ROIs with the sensor, an optical pilot signal which blinks at a slow rate of 10 Hz is introduced; the sensor can easily recognize it with a frame difference method, as shown in Fig. 5.3.

The readout scheme is shown in Fig. 5.4. The feature is based on a multiple interleaved readout of ROIs; each ROI is read out multiple times during one frame of normal images so that the ROIs can be read out much faster than the frame rate of the normal images. To explain the readout scheme simply, 6 × 6 pixels are depicted in Fig. 5.4, where two IDs exist and thus two ROIs are shown. In the figure, the number in each pixel indicates the readout order and the number in parentheses indicates the pixel involved with the ID signals. In this case, 3 ROIs images/ROI are read during one frame rate of whole images.

The frame rate of the ROIs images is actually 1.1 kfps in the sensor. The number of pixels is 320 × 240 [pixel], the ROI size is 5 × 5 [pixel], the number of IDs is 7, and the frame rate of the whole images is 30 [fps]. In this readout scheme, the system clock speed is the same as that for a conventional image sensor operating at 60 fps, and thus it consumes very little power even when reading ROIs at a high-speed frame rate. The power consumption can be further suppressed without supplying power to the column involving the ROIs.

Figure 5.5 shows a block diagram of the sensor. The sensor is operated with an ID map table, which is a 1-bit memory array with the ID positions memorized for reading out ROIs at high-speed and cutting off the power supply to pixels outside the ROIs. The pixel circuit is simple; it has only one additional transistor for column

1	2	3	4	7	8	9	10
13	14	15	16	19	20	21	22
25	26	27 ROI 1	28	31	32	33	34
37	38	(5), (29), 39, (53)	(6), (30), 40, (54)	43	44 ROI 2	45	46
49	50	(11), (35), 51, (59)	(12), (36), 52, (60)	55	(17), (41), 56, (65)	(18), (42), 57, (66)	58
61	62	63	64	67	(23), (47), 68, (71)	(24), (48), 69, (72)	70

FIGURE 5.4

Illustration of fast readout of multiple ROIs. Only 6 × 6 pixels are depicted.

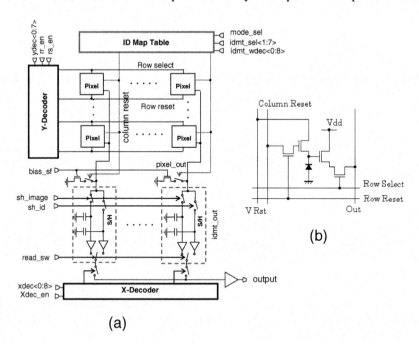

(a)

FIGURE 5.5

(a) Block diagram of a smart CMOS image sensor for fast readout of multiple ROIs. (b) Pixel circuits.

reset compared with a conventional 3T-APS, as shown in Fig. 5.5(b). It is noted that an additional transistor must be inserted between the reset line and the gate of the reset transistor, as described in Sec. 2.6.1. This transistor is used for XY-address reset or random access as described in Sec. 2.6.1 to read out pixels only in ROIs. A timing chart for the sensor is shown in Fig. 5.6. In this timing chart, fragmented normal images and ROIs images form interleaved readouts.

Figure 5.7 shows a microphotograph of the sensor. The specifications are summarized in Table 5.2. Figure 5.8(a) shows a captured normal image at 30 fps and Figs. 5.8(b) and (c) show experimental results for ID detection. ID signals are transmitted by differential 8-bit code modulated at 500 Hz from three LED modules. 36-frame ROIs images per ID, which consists of 5 × 5 pixels, are captured for detecting ID

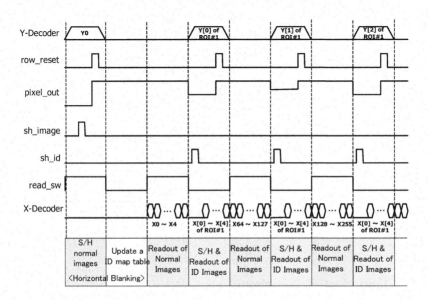

FIGURE 5.6
Timing chart for high-speed readout of multiple ROIs.

TABLE 5.2
Specifications of a smart CMOS image sensor for optical IDs tag

Technology	0.35-μm CMOS (2-poly 3-metal)
Number of pixels	QVGA (320 × 240)
Chip size	4.2 × 5.9 mm^2
Pixel size	7.5 × 7.5 μm^2
Dynamic range	54 dB
Frame rate for normal image	30 fps
Number of IDs	Max. 7
Frame rate for ID image	1.2 kfps/ID
Power consumption	3.6 mW @ 3.3 V

signals while one frame of a whole image is captured, as shown in Fig. 5.8(c). The patterns of the ROIs images successfully demonstrate the detection of each ID. These results demonstrate the image sensor can capture QVGA images at 30 fps, capturing images of three IDs at 1.1 kfps per ID. The power consumption of the sensor is 3.6 mW at a power supply of 3.3 V.

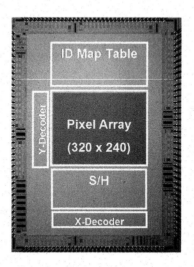

FIGURE 5.7
Microphotograph of a smart CMOS image sensor for optical ID tags.

5.2.2   Optical wireless communication

In this section, optical wireless communication or free-space optical communication using smart CMOS image sensors is introduced. Optical wireless communication has advantages over conventional communication using optical fibers and RF for the following reasons. First, setting up an optical wireless system requires only a small investment. This means such a system could be used for communication between buildings. Second, it has the potential of high-speed communication at speeds in the Gbps region. Third, it is not greatly effected by interference from electronic equipment, which is important for use in hospitals. This may be a critical issue for people implanted with an electronic medical device such as a cardiac pacemaker. Finally, it is secure because of its narrow diversity.

Optical wireless communication systems are classified into three categories from the viewpoint of usage. First are optical wireless communications for use outdoors or over long distances over 10 m. The main application target is for LANs across buildings. Optical beams in free space can easily connect two points of a building with a

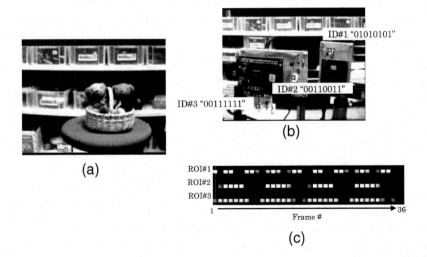

FIGURE 5.8

Images taken by a smart CMOS image sensor for optical ID tags. (a) Normal image,
(b) ID detection, (c) 36-frame ROI images/ID (in one frame for a normal image).

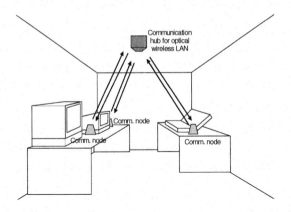

FIGURE 5.9

Example indoor optical wireless communication system. A hub station is installed in
a ceiling and several node stations are located near computers to communicate with
the hub.

fast data rate. LAN in a factory is another application for this system; the feature of
low electro-magnetic interference (EMI) as well as easy installation are suitable for
use in factories in noisy environments. Many products have been commercialized
for these applications.

The second system is for near-distance optical communication. IrDA and its relatives belongs to this category [432]. The data rate is not fast, because of the speed of LEDs. This system competes with RF wireless communication such as Bluetooth.

The third system is indoor optical wireless communication, which is similar to the second category but is intended for use as a LAN, that is, with a data rate of at least over 10 MHz. Such indoor optical wireless communication systems has been already commercialized but are limited [433]. Figure 5.9 shows an optical wireless communication system for indoor use. The system consists of a hub station installed in a ceiling and multiple node stations located near computers. It has a one-to-many communication architecture, while outdoor systems usually have one-to-one communication architecture. This one-to-many architecture causes several issues, namely, a node is required to find the hub. To achieve this task, a node is equipped with a mechanism to move the bulky optics with a photoreceiver, which means the node station is relatively large in size. The size of the hub and node is critical when the system is intended for indoor use.

Recent dramatic improvements in LEDs, including blue and white LEDs, have opened possibilities for a new type of indoor optical wireless communication called visible light communication [409]. Many applications using this concept have been proposed and developed; for example, white LED light in a room can be utilized as an transceiver, and LED lamps of automobiles can be used for communication with other automobiles. These applications are strongly related with optical ID tags, which are discussed in Sec. 5.2.1.

In optical wireless communications, it is effective to use two-dimensional detector arrays to enhance the receiver efficiency [434,435]. Replacing such two-dimensional detector arrays with a smart CMOS image sensor dedicated to fast optical wireless communication has some advantages. Such smart CMOS image sensors have been proposed and demonstrated by the Nara Institute of Science & Technology (NAIST) [436–446] and by Univ. California at Berkeley [447]. The group of Univ. California at Berkeley is working on communication between small unmanned aerial vehicles [447,448], that is, for an outdoor communication over relatively long distances compared with indoor communication.

In the following, a new scheme of indoor optical wireless LAN is introduced and described in detail, in which an image sensor based photo receiver is utilized.

Optical wireless LANs using a smart CMOS image sensor　In a new scheme for an indoor optical wireless LAN, a smart CMOS image sensor is utilized as a photoreceiver as well as a two-dimensional position-sensing device for detecting the positions of communication modules of nodes or a hub. In contrast, in a conventional system, one or more photodiodes are utilized to receive optical signals.

Figure 5.10 compares the scheme of the proposed indoor optical wireless LAN system with a conventional system. In the optical wireless LAN, optical signals must be transmitted toward the counterpart accurately to achieve a certain optical input power incident on the photo detector. The position detection of the communication modules and alignment of the light are thus significant. In a conventional optical wireless LAN system with automatic node detection, as shown in Fig. 5.10(a), a me-

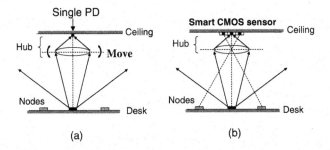

FIGURE 5.10
Optical wireless LAN systems of (a) a conventional indoor optical wireless LAN system and (b) a proposed system using a smart CMOS image sensor.

chanical scanning system for the photodetection optics is implemented to search for the hub at the node. However, the volume of the scanning system becomes bulky because the diameter of the focusing lens must be large to gather enough optical power. On the other hand, as shown in Fig. 5.10(b), using a smart image sensor as a photoreceiver is proposed as mentioned before. This approach includes several excellent features for optical wireless LAN systems. Because an image sensor can capture the surroundings of a communication module, it is easy to specify the positions of the other modules by using simple image recognition algorithms without any mechanical components. In addition, image sensors inherently capture multiple optical signals in parallel by a huge number of micro photodiodes. Different modules are detected by independent pixels when the image sensor has sufficient spatial resolution.

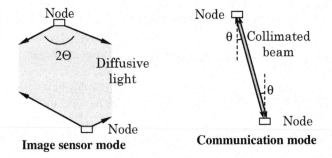

FIGURE 5.11
Two modes in the sensor, (a) image sensor (IS) mode and (b) communication (COM) mode.

The sensor has two functional modes: image sensor (IS) mode and communication

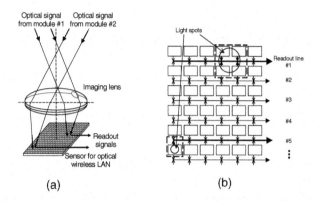

FIGURE 5.12

Illustration of space division multiplexing using a sensor with focused readout, (a) readout of multiple light spots and (b) schematic of focused readout on the sensor.

(COM) mode. Both the hub and node work in the IS mode, as shown in Fig. 5.11(a). They transmit diffusive light to notify their existence to each other. To cover the large area where the counterpart possibly exists, diffusive light with a wide radiation angle 2Θ is utilized. Because the optical power detected at the counterpart is generally very weak, it is effective to detect it in the IS mode. It is noted that, in a conventional system, the optical transceivers need to scan the other transceivers in the room by swinging bulky optics, which consumes time and energy.

After the positions are specified in the IS mode, the functional mode of the sensor for both the node and hub is switched to the COM mode. As shown in Fig. 5.11(b), they now emit a narrow collimated beam carrying communication data in the detected direction toward the counterpart. In the COM mode, photocurrents are directly read out without temporal integration from the specific pixels receiving the optical signals. Use of a collimated beam reduces power consumption at the communication module and receiver circuit area, because the output power at the emitter and the gain of the photo receiver can be reduced.

Systems using a smart CMOS image sensor have a further benefit in that they can employ space division multiplexing (SDM) and thus increase the bandwidth of the communication. Figure 5.12 illustrates SDM in a system. When optical signals from the different communication modules are received by independent pixels and read out to separate readout lines, concurrent data acquisition from multiple modules can be achieved. Consequently, the total bandwidth of the downlink at the hub can be increased in proportion to the number of readout lines.

To read out the photo current in the COM mode, a so called focused readout is utilized. As shown in Fig. 5.12(b), a light signal from a communication module is received by one or a few pixels, because the focused light spot has a finite size. The amplified photocurrents from the pixels receiving the optical signals are summed to the same readout line prepared for the COM mode, so that the signal level is not reduced.

Implementation of a smart CMOS sensor   A block diagram of a sensor is shown in Fig. 5.13(a). The sensor has one image output and four data outputs. The image is read out through a sample and hold (S/H) circuit. The chip has four independent data output channels for the COM mode. Figure 5.13(b) shows the pixel circuit, which consists of a 3T-APS, a transimpedance amplifier (TIA), digital circuitry for mode switching, and a latch memory. To select the COM/IS mode, a HIGH/LOW signal is written in the latch memory. The output from the TIA is converted to a current signal to sum the signal from the neighboring pixels. As shown in Fig. 5.13(b), each pixel has two data-output lines, on the left and right, for focused readout as mentioned above. The sum of the current signal flows is input into a column line and converted with a TIA and amplified with a main amplifier. The sensor is fabricated

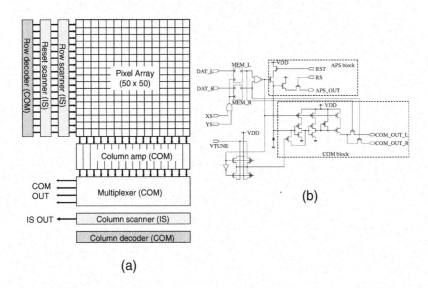

(a)

FIGURE 5.13
(a) Block diagram of the sensor. (b) Pixel circuits.

in standard 0.35-μm CMOS technology. The specifications are summarized in Table 5.3. A microphotograph of the fabricated chip is shown in 5.14.

Figure 5.15 shows experimental results for imaging and communication using the fabricated sensor. The photosensitivity at 830 nm is 70 V/(s·mW). The received waveform is shown in Fig. 5.15. The eye was open with 50 Mbps at 650 nm and 30 Mbps at 830 nm. In this sensor, an intense laser light is incident for fast communication on a few pixels, while the other pixels operate in image sensor mode. The laser beam incident on the sensor produces a large number of photocarriers, which

TABLE 5.3
Specifications of a smart CMOS image sensor for an indoor optical wireless LAN.

Technology	0.35-μm CMOS (2-poly 3-metal)
Number of pixels	50 × 50
Chip size	4.9 × 4.9 mm^2
Pixel size	60 × 60 μm^2
Photodiode	N-well/P-substrate junction
Fill factor	16%

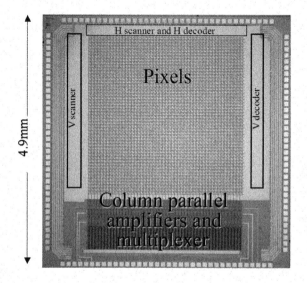

FIGURE 5.14
Microphotograph of a smart CMOS image sensor for an indoor optical wireless LAN.

travel a long distance, according to the wavelength. Some of the photo-generated carriers enter the photodiodes and affect the image. Figure 5.16 shows experimental results of effective diffusion length measured in this sensor for two wavelengths. As expected, the longer wavelength has a longer diffusion length, as discussed in Sec. 2.2.2.

Further study is required to increase the communication data rate. Preliminary results using 0.8-μm BiCMOS technology show that a data rate of 400 Mpbs/channel can be obtained. Also, a system introducing wavelength division multiplexing (WD-M) has been developed [444–446], which can increase the data rate.

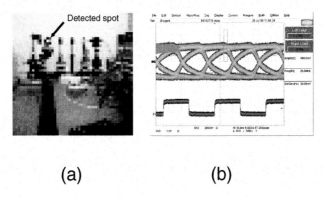

(a)   (b)

FIGURE 5.15
Experimental results of the smart CMOS image sensor. (a) Captured image, (b) 30-Mbps eye pattern for a wavelength of 830 nm.

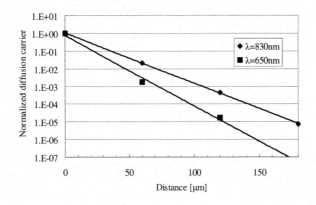

FIGURE 5.16
Measured amount of diffusion carriers on the pixels.

5.3 Biotechnology applications

In this section, applications of smart CMOS image sensors to bio-technologies are introduced. Fluorescence detection, a widely used measurement in biotechnology conventionally performed by a CCD camera installed in an optical microscope system, has been identified as an application that could be efficiently preformed by smart CMOS image sensors. Introducing smart CMOS image sensors into biotechnology would bring the benefits of integrated functions and miniaturization. These advantages are especially enhanced when combined with on-chip detection [278,449–453].

On-chip detection means that a specimen can be placed directly on the chip surface and measured. Such a configuration would make it easy to access a specimen directly so that parameters such as fluorescence, potential, pH [452], and electrochemical parameters can be measured. Integration functions are important features of a smart CMOS image sensor. These functions realize not only high SNR measurements but also functional measurements. For example, electrical simulation can be integrated into a sensor so that fluorescence caused by cell stimulation is possible; this is an on-chip electro-physiological measurement. Through miniaturization the total size of the sensing system can be reduced. Such miniaturization makes it possible to implant the sensing system. Also, on-field measurements are possible because the reduced size of the total system enhances its mobility.

Figure 5.17 illustrates typical examples of on-chip detection using smart CMOS sensors. In Fig. 5.17(a), neurons are dyed; when the neurons are excited by external stimulation, the dye emits fluorescence. The sensor has electrical simulators on the chip so that is can stimulate neurons and detect the associated fluorescence. Figure 5.17(b) shows DNA identification using a smart CMOS image sensor. DNA identification by a DNA array also utilizes fluorescence; single strands of DNA are fixed on the surface of the sensor and the target DNA with fluorescence dyes hybridize with complementary strands on the sensor surface. A third example is implantation of a smart CMOS sensor in a mouse brain, as shown in Fig. 5.17(c). In this case, the sensor is made small enough to be inserted into a mouse brain. Inside the brain, the sensor can detect fluorescence as well as stimulate neurons around it. In some cases, the intensity of fluorescence is so weak that a low light imaging technique is required, such as pulse modulation (see Sec. 3.4) [199, 200] and an APD array made of CMOS technology (see Sec. 2.3.4) [454–456].

This section introduces two example systems that exhibit the advantages described above. The first is a multi-modal image sensor, which takes electrostatic images or electrochemical images as well as performing optical imaging [457–459]. The second is an *in vivo** CMOS image sensor [460–463].

5.3.1   Smart CMOS image sensor with multi-modal functions

In this section, smart CMOS image sensors with multi-modal functions are introduced. Multi-modal functions are particularly effective for biotechnology. For example, DNA identification is more accurate correct if it is combined with an optical image with other physical values such as an electric potential image. Two examples are introduced: optical-potential multiple imaging and optical-electrochemical imaging.

---

*In vivo means "within a living organism." Similarly, *in vitro* means "in an artificial environment outside a living organism."

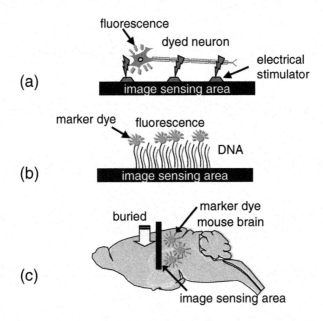

FIGURE 5.17
Conceptual illustration of on-chip detection using smart CMOS image sensors. (a) Activities of neurons are detected by fluorescence. (b) DNA array. (c) Sensor is inserted into a mouse brain for imaging brain activity.

5.3.1.1 Smart CMOS sensor for optical and potential imaging

Design of sensor A microphotograph of a fabricated sensor is shown in Fig. 5.18. The sensor has a QCIF (176 × 144) pixel array consisting of alternatively aligned 88 × 144 optical sensing pixels and 88 × 144 potential sensing pixels. The size of the pixels is 7.5 μm × 7.5 μm. The sensor was fabricated in 0.35-μm 2-poly, 4-metal standard CMOS technology.

Figure 5.19 shows circuits for a light-sensing pixel, a potential-sensing pixel, and the column unit. A potential-sensing pixel consists of a sensing electrode, a source follower amplifier, and a select transistor. The sensing electrode is designed with a top metal layer and is covered with a passivation layer of LSI, that is silicon nitride (Si_3N_4). The sensing electrode is capacitively coupled with the potential at the chip surface. Using the capacitive coupling measurement method, no current flows from the image sensor and perturbation caused by a measurement is smaller than that caused by a conductive coupling sensor system, such as a multiple electrode array.

Experimental results Figure 5.20 shows an as-captured (optical and potential) image and reconstructed images. The sensor is molded from silicone rubber and only a part of the sensor array is exposed to the saline solution in which the sensor is im-

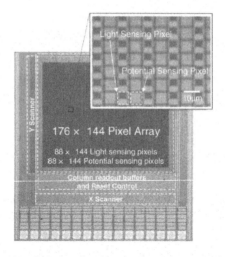

FIGURE 5.18
Microphotographs of a fabricated smart CMOS image sensor for optical and potential dual imaging.

mersed. A voltage source controls the potential of the saline solution via a Ag/AgCl electrode dipped in the solution. As shown in Fig. 5.20, the as-captured image is complicated because optical and potential images are superimposed in the image. However, the data can be divided into two different images. The dust and scratches observed in the microphotograph of Fig. 5.20(a) and a shadow of the Ag/AgCl electrode are clearly observed in the optical image as shown in Fig. 5.20(b). On the other hand, the potential at the exposed region shows a clear contrast to the covered region in the potential image in Fig. 5.20(c). As described later, the potential-sensing pixel shows a pixel-dependent offset caused by trapped charge in the sensing electrode. However, the offsets are effectively canceled in the image reconstruction process. In the case of a sine wave between 0 and 3.3 V (0.2 Hz) applied to the saline solution, the potential change in the solution can be clearly observed in the region that is exposed to the solution. No crosstalk signal was observed in either the covered potential-sensing pixels or the optical image. Simultaneous optical and potential imaging was successfully achieved.

Figure 5.21 shows potential imaging results using conductive gel probes. Two conductive gel probes were placed on the sensor and the voltages of the gels were controlled independently. The images clearly show the voltages applied to the gel spots. The difference in applied voltage was captured with good contrast. The sensor is capable of taking not only still images but also moving images over 1—10 fps. The frame rate will be improved in the next version of the sensor. The resolution is currently smaller than 6.6 mV, which is sufficient to detect DNA hybridization [464]. It is expected that the potential resolution will be improved to the 10-μV level. Since

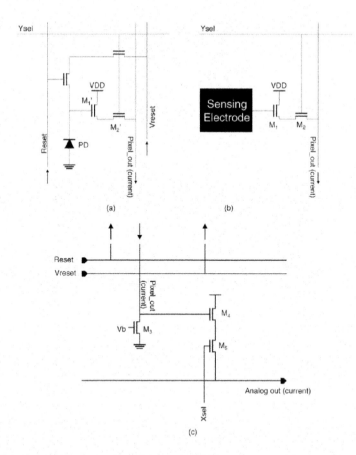

(a)   (b)

(c)

FIGURE 5.19

Circuits of a smart CMOS image sensor for optical and potential dual imaging for (a) a light-sensing pixel, (b) a potential-sensing pixel, and (c) a column unit.

the present sensor does not have an on-chip analog to digital converter (ADC), the data suffers from noise introduced into the signal line between the sensor and ADC chip. On-chip ADC circuitry is required to use the image sensor for high-resolution neural recording.

5.3.2 Potential imaging combining MEMS technology

P. Abshire and coworkers at Univ. Maryland have developed a potential imaging device using CMOS technology combined with micro-electro-mechanical system (MEMS) technology [465]. The device is illustrated in Fig. 5.22. Microvials, small holes for individually culturing cells, are fabricated on the surface of a CMOS LSI

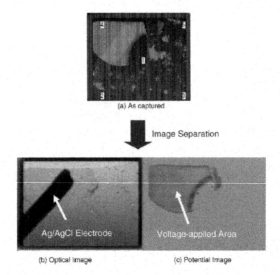

FIGURE 5.20

Images taken by the sensor: (a) as-captured image, (b) (c) reconstructed optical images from image (a), (c) reconstructed potential image from image (a).

chip integrated with an array of potential-sensing circuits. MEMS technology is used to fabricate the microvials with controllable lids that can be opened and closed by an integrated actuator. By using this device, cells are introduced by dielectrophoresis and tested over a long time period. The device is now under development and has been used to demonstrate applications of smart CMOS image sensors combining MEMS technology in bio- and medical technologies.

5.3.3 Smart CMOS sensor for optical and electrochemical imaging

Usually, fluorescence is used to detect hybridized target DNA fragments on probe D-NA spots, as described in the begining of this section. Electrochemical measurement is another promising detection scheme that could be an alternative or a supplementary method for micro-array technology [466–468]. Various electrochemical methods to detect hybridized biomolecules have been proposed and some are being used in commercial equipment. Some groups have reported LSI-based sensors for on-chip detection of biomolecules using electrochemical detection techniques [459,467,469].

Design of sensor   Figure 5.23 shows microphotographs of a smart CMOS image sensor for optical and electrochemical dual-imaging. The sensor was fabricated in 0.35-μm, 2-poly, 4-metal standard CMOS technology. It consists of a combined optical and electrochemical pixel array and control/readout circuitry for each function. The combined pixel array is a 128×128 light-sensing pixel array, partly replaced

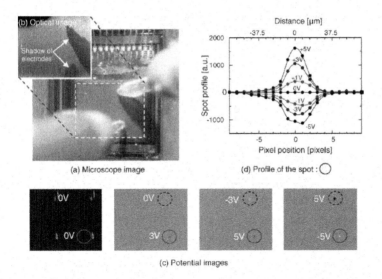

(a) Microscope image

(d) Profile of the spot : ◯

(c) Potential images

FIGURE 5.21

Experimental setup and results of potential imaging. (a) Microphotograph of the measurement. Two gel-coated probes are placed on the sensor surface and potentials are applied. (b) On-chip optical image and (c) on-chip potential images. The profiles of the spots indicated by the solid circles in (c) are shown in (d).

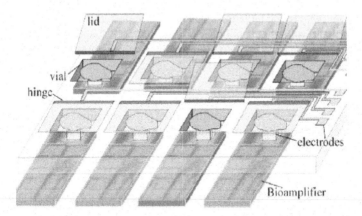

FIGURE 5.22

Illustration of CMOS-MEMS based biochip [465]. Each pixel consists of a microvial and an amplifier to sense extracellular potentials. Courtesy of Prof. P. Abshire at Univ. Maryland.

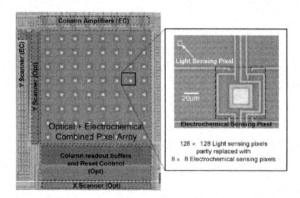

FIGURE 5.23
Microphotograph of a fabricated smart CMOS sensor for optical and electrochemical
dual imaging.

with electrochemical sensing pixels. The light-sensing pixel is a modified 3T-APS
with a pixel size of 7.5 μm× 7.5 μm. The electrochemical-sensing pixel consists of
an exposed electrode with an area of 30.5 μm× 30.5 μm using a transmission-gate
switch for row selection. The size of the electrochemical-sensing pixel is 60 μm×
60 μm. Thus, 8 × 8 light-sensing pixels are replaced by one electrochemical-sensing
pixel. The sensor has an 8 × 8 electrochemical pixel array embedded in the optical
image sensor. Due to the large mismatch in operating speed between the optical image sensor and electrochemical image sensor, the optical and electrochemical pixel
arrays are designed to work independently. Figure 5.24 shows the schematics of the
sensor.

A voltage-controlled current measurement approach can be used for electrochemical measurements for on-chip bimolecular micro-array technology. Options include
cyclic voltammetry (CV) [468] and differential pulse voltammetry [466], which have
been reported to be feasible for detecting hybridized DNA. A current-sensing voltage
follower is used for on-chip, multisite electrochemical measurements. By inserting a
resistance in the feedback path of the voltage follower (unity-gain buffer), the circuitry can perform voltage-controlled current measurements. This circuit configuration
has been widely used in electrochemical potentiostats and patch-clamp amplifiers.

Experimental results   Two-dimensional (2D) arrayed CV measurements have been
performed and 8 × 8 CV curves were obtained using a single frame measurement.
For on-chip measurements, gold was formed as electrodes on the exposed aluminum
electrodes of the electrochemical sensing pixels. Because of its chemical stability
and affinity to sulfur bases, gold has been a standard electrode material for electrochemical molecular measurements. Au/Cr (300 nm/10 nm) layers were evaporated
and patterned into the 30.5 μm × 30.5 μm electrochemical sensing electrodes. The

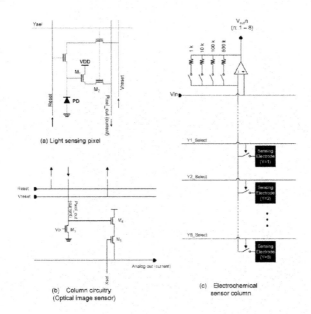

FIGURE 5.24
Pixel circuits for (a) optical sensing and (b) electrochemical sensing and (c) column circuits.

sensor was then mounted on a ceramic package and connected with aluminum wires. The sensor with connecting wires was molded with an epoxy rubber layer. Only the combined pixel array was kept uncovered and exposed to the measurement solution.

A two-electrode configuration was used for the arrayed CV measurements. An Ag/AgCl electrode was used as the counter electrode. The work electrode was an 8×8-array gold electrode. As a model subject for the 2D CV measurement, an agarose gel island with a high resistivity in a saline solution was used. 8×8 CV curves were measured to take images of the electrochemical characteristics. The potential of the Ag/AgCl counter electrode was cyclically scanned between -3 and 5 V for each electrochemical row with a scan speed of 1 Hz. Figure 5.25 shows the results of the arrayed CV measurement. The observed CV profiles show different features depending on the situations of the individual measurement electrodes.

5.3.4 Fluorescence detection

Fluorescence contains important information in biotechnology measurements. In fluorescence measurements, excitation light is typically used, and thus fluorescence can be distinguished as a signal light from the background signal due to the excitation light. To suppress the background light, on-chip color filters for signal rejection [460–463] and potential profile control [266, 267] have been proposed and

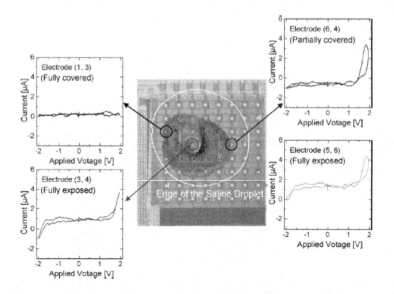

FIGURE 5.25
Experimental results of 2D arrayed CV measurements.

demonstrated. Potential profile control is explained in Sec. 3.7.3. In the next section, a smart CMOS image sensor for *in vivo* mouse brain imaging is presented.

5.3.4.1 Smart CMOS image sensor for *in vivo* mouse brain imaging

An important application is the imaging of the brain to study its learning and memory functions [470]. Current technology for imaging the brain requires expensive equipment that has limitations in terms of image resolution and speed or imaging depth, which are essential for the study of the brain [471]. As shown in Fig. 5.26, a miniaturized smart CMOS image sensor (denoted "CMOS sensor" in the following) is capable of real time *in vivo* imaging of the brain at arbitrary depths.

Image sensor implementation   An image sensor has been fabricated in standard 0.35-μm CMOS process. The image sensor is based on a 3T-APS with a modification for pulse width modulation (PWM) output, as shown in Fig. 5.27 with a chip microphotograph. The sensor consists of a digital and analog output for interface with an external read-out circuit. Image sensing at near video rates is performed via the analog output. The digital output, which enables PWM output, is suitable where a long integration time is required in static imaging. The PWM photosensor is described in Sec. 3.4.1. It was designed to be large enough to image a mouse hippocampus yet small enough for invasive imaging of each brain hemisphere independently. Its specifications are listed in Table 5.4. It is noted that this sensor has

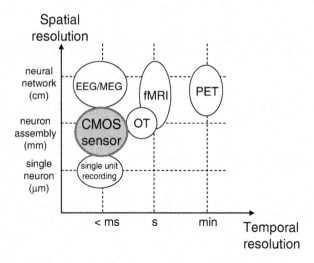

FIGURE 5.26

Current neural imaging technology. EEG: electroencephalography, MEG: magnetoencephalography, OT: optical topography, fMRI: functional magnetic resonance imaging, PET: positron emission tomography.

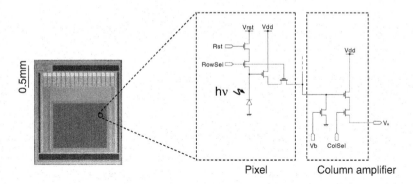

FIGURE 5.27

Pixel circuits of a smart CMOS image sensor for *in vivo* imaging. The right figure shows a microphotograph of the fabricated chip.

only an imaging function, though it is possible to integrate the sensor with electrical stimulation functions.

Packaging is a critical issue for applications where the sensor is inserted deeply into the mouse brain. Figure 5.28 illustrates a packaged device; it is integrated with a smart CMOS image sensor and a UV-LED for excitation on a flexible polyimide substrate. A dedicated fabrication process was developed, which makes it possible

TABLE 5.4

Specifications of a smart CMOS image sensor for *in vivo* imaging

Technology	0.35-μm CMOS (2-poly 4-metal)
Number of pixels	176 $\times$ 144 (QCIF)
Chip size	2 $\times$ 2.2 mm^2
Pixel size	7.5 $\times$ 7.5 μm^2
Photodiode	N-well/P-substrate junction
Fill factor	29%
Pixel type	Modified 3T-APS

to realize an extremely compact device measuring 350 μm in thickness.

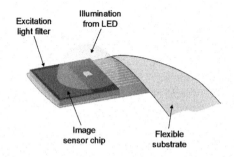

FIGURE 5.28

Conceptual illustration of an *in vivo* imaging device.

The fabrication process is as follows. The chip is first thinned down to about 200 μm. It is attached to a flexible and bio-compatible polyimide substrate and protected with a layer of transparent epoxy. A filter with high selectivity for fluorescence e-mission of 7-amino-4-methylcoumarin (AMC) is spin-coated onto the surface of the device. A transmittance of -44 dB at 450 nm, where the fluorescence is emitted, was achieved by this simple method. The transmittance obtained is comparable to discrete filters used in fluorescence microscopes and is sufficient for on-chip fluorescence imaging. A chip LED with UV emission (365 nm) was then attached on the sensor. Finally, to demonstrate the operation of the CMOS sensor device for imaging brain activity, the device was further developed to include a needle for injecting a fluorophore substrate and an excitation light fiber. The device is interfaced to a PC to readout the output signal as well as input control signals. Figure 5.29 shows the fully developed device.

Using this device in conjunction with a fluorophore substrate (PGR-MCA*, VPR-

---

*Pyr-Gly-Arg-4methyl-coumarin

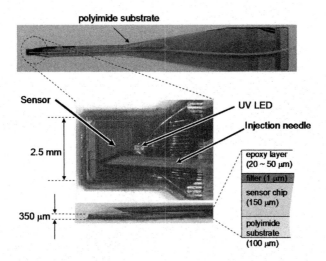

FIGURE 5.29
Photograph of a fabricated *in vivo* imaging device, including a close-up photograph around the sensor.

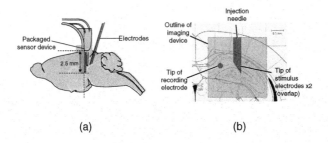

(a)   (b)

FIGURE 5.30
Experimental setup with device inserted in a mouse brain. (a) Sagittal plane view of device inside mouse brain. (b) Coronal plane view of device inside mouse brain.

MCA*), experiments were performed to detect the presence of serine protease inside a mouse hippocampus, such as neuropsin and tPA. VPR-MCA was used to detect activated neuropsin [472], while PGR-MCA specifically targeted tPA. In the experiment, kainic acid (KA) was introduced as an agent, which causes serine protease to be expressed extracellularly from postsynaptic neuronal cells. The experimental setup with the device inserted is shown in Fig. 5.30.

---

*Boc-Val-Pro-Arg-4-methyl-coumarine

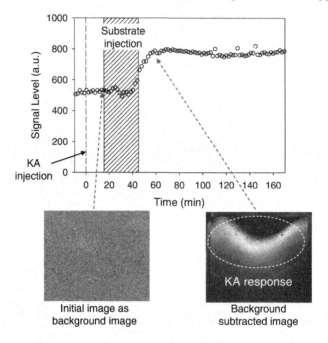

FIGURE 5.31
Experimental results of *in vivo* deep brain imaging using a smart CMOS image sensor.

Experimental results for *in vivo* imaging   In the experiment, serine protease activity was observed in real time by imaging AMC fluorescence. The AMC fluorophore is released from the substrate due to the presence of the serine protease. Multiple pixel locations of the image sensor near the outlet of the injection needle were selected and plotted. A plot of the signal level from a single location is shown in Fig. 5.31. From the result, a significant increase in fluorescence signal is observed at about 1 hr after KA injection. This increase in the signal is the result of an increase in serine protease activity localized near the injection needle. In order to confirm this observation, the mouse brain was extracted at the end of the experiment, sliced, and observed under a fluorescence microscope.

This experiment successfully demonstrated the capability of the CMOS imaging device for detecting functional brain activity in real time. In addition, by using the device for *in vivo* serine protease imaging, the experiment independently verified reported findings on the effect of KA on the hippocampus.

Another experiment shows that minimal injury was inflicted on the brain using this method as the brain continued to function and respond normally with the device inside it.

In the next step, stimulus electrodes are integrated on a chip. Figure 5.32 shows an *in vivo* smart CMOS image sensor with a UV-LED excitation source. Several holes

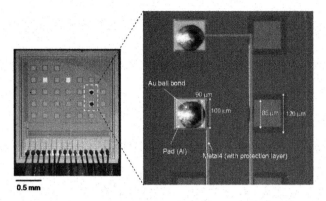

FIGURE 5.32
Advanced smart CMOS image sensor for *in vivo* imaging integrated with stimulus electrodes.

through the chip were fabricated for the illumination of the UV-LED. These holes may also be used for the injection of KA.

5.4 Medical applications

In this section, two topics of applications for smart CMOS image sensors are presented: capsule endoscopes and retina prosthesis. Smart CMOS image sensors are suitable for medical applications for the following reasons. First, they can be integrated with signal processing, RF, and other electronics, that is, a system-on-chip (SoC) is possible. This can be applied to a capsule endoscope, which requires a system with small volume and low power consumption. Second, smart functions are effective for medical use. Retinal prosthesis is such an example requiring an electronic stimulating function on a chip. In the near future, the medical field will be one of the most important applications for smart CMOS image sensors.

5.4.1 Capsule endoscope

An endoscope is a medical instrument for observing and diagnosing organs such as the stomach and intestines by being inserting into the body. It is employed with a C-CD camera with a light-guided glass fiber to illuminate the area being observed. An endoscope or push-type endoscope is a highly integrated instrument with a camera, light guide, small forceps to pick up tissue, a tube for injecting water to clean tissues, and an air tube for enlarging affected regions. A capsule endoscope is a kind of endoscope developed in 2000 by Given Imaging in Israel [473]. It is currently available for sale in the US and Europe. Olympus has also developed a capsule endoscope, which is for sale in Europe. Figure 5.33 shows a photograph of Olympus's capsule endoscope.

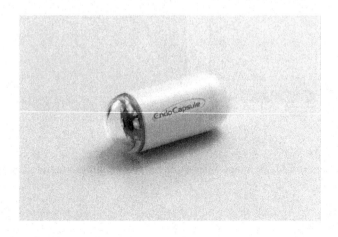

FIGURE 5.33
Capsule endoscope. A capsule with the length of 26 mm and a diameter of 11 mm is used with a dome, optics, white LEDs for illumination, CCD camera, battery, RF electronics, and antenna. Courtesy of Olympus.

A capsule endoscope uses an image sensor, imaging optics, LEDs for illumination, RF circuits, antenna, battery, and other elements. A user swallows a capsule endoscope and it automatically moves along the digestive organs. Compared with a conventional endoscope, a capsule endoscope causes a user less pain. It is noted that a capsule endoscope is limited to use in the small intestine, and is not used for the stomach or large intestine. Recently, a capsule endoscope for observing the esophagus is developed by Given Imaging [474]. It has two cameras to image forward and rear scenes.

Smart CMOS image sensor for capsule endoscope A capsule endoscope is a kind of implanted device, and hence the critical issues are size and power consumption. A smart CMOS image sensor is thus suitable for this purpose. When applying a CMOS image sensor, color realization must be considered. As discussed in Sec. 2.11, a CMOS image sensor uses a rolling shutter mechanism. Medical use generally requires color reproducibility so that three image sensors or three light sources are preferable, as discussed in Sec. . In fact, conventional endoscopes use the three-light sources method. For installing a camera system in a small volume, the three-light sources method is particularly suitable for a capsule endoscope. However, due to the rolling shutter mechanism, the three-light sources method cannot be applied to

CMOS image sensors. In a rolling shutter, the shutter timing is different in every row, and hence the three-light sources method where each light emits at different timing can not be applied. Present commercially available capsule endoscopes use on-chip color filters either in a CMOS image sensor (Given Imaging) or CCD image sensor (Olympus). To apply the three-light sources method in a CMOS image sensor, a global shutter is required. Another method has been proposed that calculates the color reproducibility. As a RGB mixing ratio is known a priori in a rolling shutter when using the three-light source method, RGB can be separated by calculating outside the chip [475]. Although color reproducibility must be evaluated in detail, it is a promising method for CMOS image sensors with three-light sources method.

Because a capsule endoscope operates by a battery, the power consumption of the total electronics should be small. Also, the total volume should be small. Thus, SoC for the imaging system including RF electronics is effective. For this purpose, SoC with a CMOS image sensor and RF electronics has been reported [476]. As shown in Fig. 5.34, the fabricated chip has only one I/O pad of a digital output besides a power supply (Vdd and GND) integrated with BPSK (Binary Phase Shift Keying) modulation electronics. The chip consumes 2.6 mW under a condition of 2 fps with QVGA format. A SoC for a capsule endoscope has been reported, though the image sensor is not integrated. This system has the capability of wirelessly transmitting data of 320×288 pixels in 2 Mbps with a power consumption of 6.2 mW [477].

Another desirable function for a capsule endoscope is on-chip image compression. There are several reports of on-chip compression [478–481] and in the near future it is expected that this function will be employed by a capsule endoscope. These SoCs will be used in capsule endoscopes as well as combining with technologies such as micro electro-mechanical systems (MEMS), micro total analysis system (μTAS), and lab-on-chip (LOB) to monitor other physical values such as potential, pH, and temperature [482, 483]. Such multi-modal sensing is suitable for the smart CMOS image sensor described in the previous section, Sec. 5.3.1.

5.4.2 Retinal prosthesis

In early work in the field, MOS image sensors have been applied for helping the blind. The Optacon, or optical-to-tactile converter, is probably the first use of a solid-state image sensor for the blind [484]. The Optacon integrated scanning and readout circuits and was compact in size [119, 485]. Retinal prosthesis is like an implantable version of the Optacon. In the Optacon, the blind perceive an object through tactile sense, while in retinal prosthesis the blind perceive an object through an electrical stimulation of vision-related cells by an implanted device.

A number of studies have been carried out considering different implantation sites [486] such as the cortical region [487,488], optic nerve [489], epi-retinal space [490–496], and sub-retinal space [497–503]. Recently another approach called supra-choroidal transretinal stimulation (STS) has been proposed and developed [504–506].

Implantation in the retinal space or ocular implantation prevents infection and can be applied to patients suffering from retinitis pigmentosa (RP) and age-related

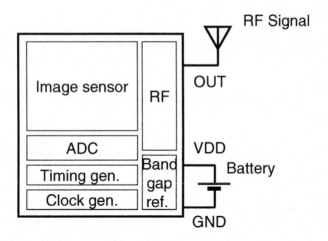

FIGURE 5.34
SoC including a CMOS image sensor for capsule endoscopes [476]. ADC: analog-to-digital converter. Timing gen.: timing pulse generator. Clock gen.: internal clock generator. A cyclic ADC is used in this chip [476].

macular degeneration (AMD) where retinal cells other than the photoreceptors still function. It is noted that both RP and AMD are diseases with no effective remedies yet. The structure of the retina is shown in Fig. C.1 of Appendix C.

While in the epi-retinal approach, ganglion cells (GCs) are stimulated, in the sub-retinal approach the stimulation is merely a replacement of the photoreceptors and thus in an implementation of this approach it is likely that bipolar cells as well as GCs will be stimulated. Consequently, the sub-retinal approach has the following advantages over the epi-retinal approach: there is little retinotopy, that is, the stimulation points correspond well to the visual sense, and it is possible to naturally utilize optomechanical functions such as the movement of the eyeball and opening and closing of the iris. Figure 5.35 illustrates the three methods of epi- and sub-retinal stimulations and STS.

Sub-retinal approach using PFM photosensor In sub-retinal implantation, a photosensor is required in order to integrate the stimulus electrode. Thus far, a simple photodiode array without any bias voltage, that is, a solar cell mode, has been used as a photosensor mainly due to its simple configuration in that there is no need for a power supply [497–499]. The photocurrent is directly used as the stimulus current into the retinal cells. In order to generate sufficient stimulus current using a photosensor in a daylight environment, a pulse frequency modulation (PFM) photosensor has been proposed for the sub-retinal approach [201, 212] and PFM-based retinal prosthesis devices have been developed as well as a simulator for STS [190, 193, 202–211, 507–511]. Recently, several groups have also devel-

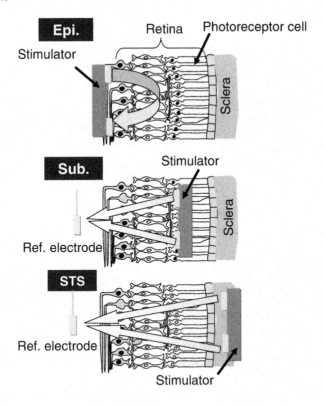

FIGURE 5.35
Three methods of retinal prosthesis for ocular implantation.

oped a PFM photosensor or pulse-based photosensor for use in sub-retinal implantation [213–216, 512–515].

PFM appears to be suitable as a retinal prosthesis device in sub-retinal implantation for the following reasons. First, PFM produces an output of pulse streams, which would be suitable for stimulating cells. In general, pulse stimulation is effective for evoking cell potentials. In addition, such a pulse form is compatible with logic circuits, which enables highly versatile functions. Second, PFM can operate at a very low voltage without decreasing the signal-to-noise ratio. This is suitable for an implantable device. Finally, its photosensitivity is sufficiently high for detection in normal lighting conditions and its dynamic range is relatively large. These characteristics are very advantageous for the replacement of photoreceptors. Although the PFM photosensor is essentially suitable for application to a retinal prosthesis device, some modifications are required and these will be described herein.

Epi-retinal approach using an image sensor device   It is noted that the sub-retinal approach is natural when using imaging with stimulation because imaging

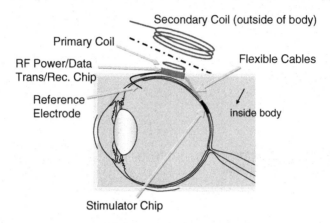

FIGURE 5.36
Total system of ocular implantation for retinal prosthesis.

can be done on the same plane of stimulation. Some epi-retinal approaches, however, can use implanted imaging device with stimulation. As mentioned in Sec. 3.6.1.2, silicon-on-sapphire (SOS) is transparent so that is can be used as an epi-retinal approach by using a back-illuminated image sensor. For the back-illuminating configuration, the imaging area and stimulation can be placed on the same plane. A PFM photosensor using SOS CMOS technology has been demonstrated for the epi-retinal approach [203].

Another epi-retinal approach using an image sensor is to use three-dimensional (3D) integration technology [213,512,513]. Figure 5.37 shows the concept of retinal prosthesis using 3D integration technology.

LSI for ocular implantation   It should be noted that there are many technical challenges to overcome when applying LSI-based simulator devices to retinal prosthesis. Although many epi-retinal approaches have utilized LSIs [492, 494, 516], the subretinal approach has difficulties in using LSI because it is completely implanted into the tissues and must work for both image sensors and electric simulators.

Bio-compatibility   An LSI-based interface must be bio-compatible. The standard LSI structure is unsuitable for a biological environment; silicon nitride is conventionally used as a protective top layer in standard LSIs, but will be damaged in a biological environment for long-time implantation.

Fabrication compatible with standard LSI   The stimulus electrodes must be compatible with the standard LSI structure. Wire-bonding pads, which are made of aluminum, are typically used as input–output interfaces in standard LSIs, but are completely inadequate as stimulus electrodes for retinal cells, because aluminum

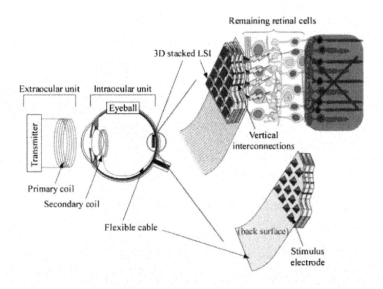

FIGURE 5.37
Epi-retinal approach using 3D integration technology [213]. Courtesy of Prof. M. Koyanagi at Tohoku Univ.

dissolves in a biological environment. Platinum is a candidate for stimulus electrode materials.

Shape of stimulus electrode In addition to electrode materials, the shape of the electrode affects the efficiency of stimulation. A convex shape is suitable for efficient stimulation, but the electrodes in LSI are flat. Thus, formation of convex-shaped platinum stimulus electrodes on an LSI is a challenge. These issues are discussed in detail in Ref. [212].

Modification of PFM photosensor for retinal cell stimulation To apply a PFM photosensor to retinal cell stimulation, the PFM photosensor must be modified. The reasons for the modification are as follows.

Current pulse The output from a PFM photosensor has the form of a voltage pulse waveform, whereas a current output is preferable for injecting charges into retinal cells, even if the contact resistances between the electrodes and the cells are changed.

Biphasic pulse Biphasic output, that is, positive and negative pulses, is preferable for charge balance in the electrical stimulation of retinal cells. For clinical use, charge balance is a critical issue because residue charges accumulate in living tissues, which may cause harmful effects to retinal cells.

Frequency limit   An output frequency limitation is needed because an excessively high frequency may cause damage to retinal cells. The output pulse frequency of the original PFM device shown in Fig. 3.11 is generally too high (approximately 1 MHz) for stimulating retinal cells. The frequency limitation, however, causes a reduction in the range of the input light intensity. This problem is alleviated by introducing a variable sensitivity wherein the output frequency is divided into 2^{-n} portions with a frequency divider. This idea is inspired by the light-adaptation mechanism in animal retina, as illustrated in Fig. C.2 of Appendix C. Note that the digital output of the PFM is suitable for the introduction of the logic function of the frequency divider.

5.4.2.1   PFM photosensor retinal simulator

Based on the above modifications, a pixel circuitry has been designed and fabricated using standard 0.6-μm CMOS technology [202]. Figure 5.38 shows a block diagram of the pixel. Frequency limitation is achieved by a low pass filter using switched capacitors. A biphasic current pulse is implemented by switching the current source and sink alternatively.

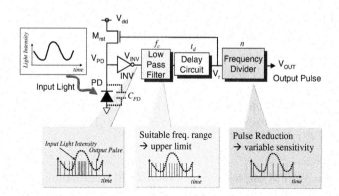

FIGURE 5.38
Block diagram of PFM photosensor pixel circuit modified for retinal cell stimulation.

Figure 5.39 shows experimental results of variable photosensitivity using the chip. The original output curve has a dynamic range of over 6-log (6th-order range of input light intensity), but is reduced to around 2-log to be limited at 250 Hz when the low pass filter is turned on. By introducing variable sensitivity, the total coverage of input light intensity becomes 5-log between $n = 0$ and $n = 7$, where n is the number of divisions.

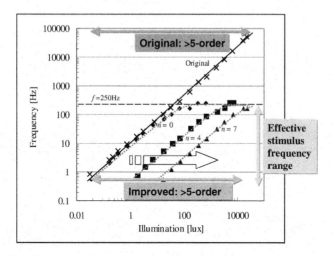

FIGURE 5.39
Experimental results of variable photosensitivity of the PFM photosensor.

Image preprocessing using the PFM photosensor   When the PFM-based simulator device is applied to a retinal prosthesis, the resolution is less than approximately 30×30. This limitation arises because the electrode pitch is larger than 100 μm according to the electro-physiological experiments, and the width of the chip is less than approximately 3 mm according to the implantation operation. In order to obtain a clear image with such a low resolution, image processing, such as edge enhancement, is necessary. In order to implement image processing, a new principle of spatial filtering in the pulse frequency domain has been developed [190] and is described below.

Spatial filtering is generally based on the spatial correlation operation using a kernel h as

$$g(x,y) = \sum_{x'}\sum_{y'} h(x',y')f(f+x',y+y'), \qquad (5.1)$$

where $f(x,y)$ and $g(x,y)$ indicate the pixel values at (x,y) of the input and output images, respectively.

Usually, f, g, and h are analog values for analog image processing or integers for digital image processing. In this scheme, f and g are represented as a pulse frequency. Thus, for this implementation, a method to represent the kernel weight in the pulse domain is developed as follows. An interaction with neighboring pixels as the gate control of the pulse stream from the neighboring pixels is introduced.

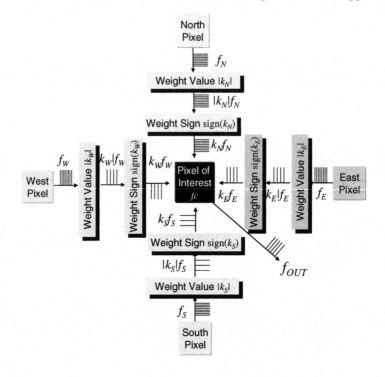

FIGURE 5.40

Conceptual illustration of image processing in the pulse frequency domain.

This concept is illustrated in Fig. 5.40. The absolute value of the kernel weight $|h|$ is expressed as the on–off frequency of the gate control. The sign is expressed as follows. In order to realize negative weights in the spatial filtering kernel, the pulses from a pixel interacts with those from its neighboring pixels to make them disappear. For positive weights, the pulses from the pixel are merged with those from the neighbors. These mechanisms can be achieved by simple digital circuitry. In the architecture, a 1-bit pulse buffering memory is used to absorb the phase mismatch between the interacting pulses. It is noted that the operation here has the nature of a stochastic process [191], and hence another architecture may be possible, as seen in biological information processing with pulse coding [188].

The proposed architecture allows the execution of fundamental types of image processing, such as edge enhancement, blurring, and edge detection. The advantage of this architecture over straightforward digital spatial filtering is that the number of required logic gates is small because there is no need for adders or multipliers.

A binarization circuit is implemented based on an asynchronous counter with N-bit D-flip flop (D-FF). The input of the D-FF of the MSB is fixed to HIGH. When $2N - 1$ pulses are input, the output of the counter turns from HIGH to LOW. This means that the counter works as a digital comparator with a fixed threshold of $2N - 1$.

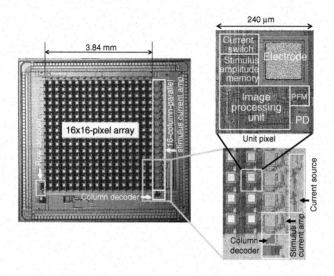

FIGURE 5.41

Microphotograph of the fabricated PFM chip for retinal prosthesis.

According to the architecture, a PFM-based retinal prosthesis device with 16 × 16 pixels is demonstrated. Figure 5.41 shows microphotographs of the fabricated chip, peripheral circuits, and its pixels. This chip implements two neighboring correlations of upper (north) and left (west) pixels. As shown in Fig. 5.41, each pixel has a PFM photosensor, the above-described image processing circuits, stimulating circuits, and a stimulus electrode, that is, this chip can stimulate retinal cells. This chip is used for *in vitro* electro-physiological experiments of stimulating retinal cells, as described in the following section. Figure 5.42(b) shows the experimental results of image processing using the chip: showing an original image, edge enhancement, and blurring. These results clearly demonstrate that the proposed architecture works properly.

Experimental results using the PFM photosensor chip with a further increase in the number of pixels and number of correlation pixels are now demonstrated. The chip has 32 × 32 pixels, in which capacitive feedback PFMs are implemented with 4-neighboring correlation processing, as shown in Fig. 5.43, although this chip has no retinal stimulation circuits [209]. The capacitive feedback PFM is described in Sec. 3.4.2.2. The experimental results are shown in Fig. 5.44, where smooth image preprocessing results are obtained.

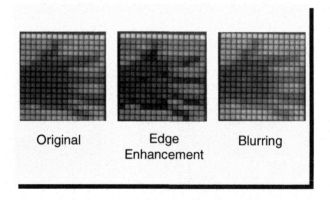

FIGURE 5.42
Image processing results using the PFM image sensor using two neighboring correlations [190].

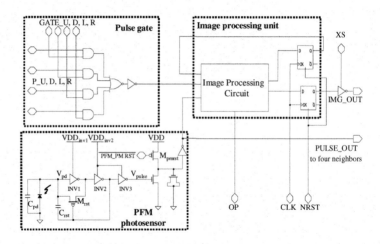

FIGURE 5.43
Pixel schematic of a capacitive feedback PFM with image preprocessing functions [209].

Application of PFM photosensor to the stimulation of retinal cells   In this section, the PFM-based simulator described in the previous section is demonstrated to be effective in stimulating retinal cells. In order to apply the Si-LSI chip to electrophysiological experiments, we must protect the chip against the biological environment, and make an effective stimulus electrode that is compatible with the standard LSI structure. In order to meet these requirements, a Pt/Au stacked bump electrode has been developed [508–511]. However, due to the limited scope of this book, this

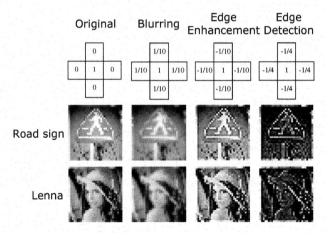

Original	Blurring	Edge Enhancement	Edge Detection

FIGURE 5.44
Image processing results using the capacitive feedback PFM image sensor using four-neighboring correlation [209].

electrode will not be described.

In order to verify the operation of the PFM photosensor chip, *in vitro* experiments using detached bullfrog retinas were performed. In this experiment, the chip acts as a stimulator that is controlled by input light intensity, as is the case in photoreceptor cells. A current source and pulse shape circuits are integrated onto the chip. The Pt/Au stacked bump electrode and chip molding processes were performed as described in [508–511].

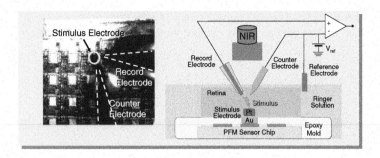

FIGURE 5.45
Experimental setup of *in vitro* stimulation using the PFM photosensor [208].

A piece of bullfrog retina was placed, with the retinal ganglion cell (RGC) side face up, on the surface of the packaged chip. Figure 5.45 shows the experimental setup. Electrical stimulation was performed using the chip at a selected single pixel. A tungsten counter electrode with a tip diameter of 5 μm was placed on the retina, producing a trans-retinal current between the counter electrode and the chip electrode. A cathodic-first biphasic current pulse was used as the stimulation. The pulse parameter is described in the inset of Fig. 5.45. Details of the experimental setup are given in Ref. [208]. Note that NIR light (near-infrared) does not excite the retinal cells, but does excite the PFM photosensor cells. Figure 5.46 also demonstrates the experimental results of evoking retinal cells with the PFM photosensor, which is illuminated by input NIR light. The firing rate increases in proportion to the input NIR light intensity. This demonstrates that the PFM photosensor activates the retinal cells through the input of NIR light, and suggests that it can be applied to human retinal prosthesis.

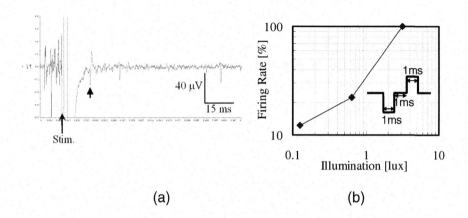

(a)   (b)

FIGURE 5.46
Experimental results of *in vitro* stimulation using the PFM photosensor. (a) Example obtained waveform obtained. (b) Firing rate as a function of input light intensity [208].

A

Tables of constants

TABLE A.1

Physical constants at 300 K [77]

Quantity	Symbol	Value	Unit
Avogadro constant	N_{AVO}	6.02204×10^{23}	mol^{-1}
Boltzmann constant	k_B	1.380658×10^{-23}	J/K
Electron charge	e	$1.60217733 \times 10^{-16}$	C
Electron mass	m_e	$9.1093897 \times 10^{-31}$	kg
Electron volt	eV	$1 \text{ eV} = 1.60217733 \times 10^{-16}$ J	
Permeability in vacuum	μ_o	1.25663×10^{-6}	H/m
Permittivity in vacuum	ε_o	$8.854187817 \times 10^{-12}$	F/m
Planck constant	h	$6.6260755 \times 10^{-34}$	J·s
Speed of light in vacuum	c	2.99792458×10^8	m/s
Thermal voltage at 300 K	$k_B T$	26	meV
Thermal noise in 1 fF capacitor	$\sqrt{k_B T / C}$	5	μV
Wavelength of 1-eV quantum	λ	1.23977	μm

TABLE A.2

Properties of some materials at 300 K [77]

Property	Unit	Si	Ge	SiO_2	Si_3N_4
Bandgap	eV	1.1242	0.664	9	5
Dielectric constant		11.9	16	3.9	7.5
Refractive index		3.44	3.97	1.46	2.05
Intrinsic carrier conc.	cm^{-3}	1.45×10^{10}	2.4×10^{13}		
Electron mobility	cm^2/Vs	1430	3600	-	-
Hole mobility	cm^2/Vs	470	1800	-	-

B

Illuminance

Figure B.1 shows typical levels of illuminance for a variety of lightning conditions. The absolute threshold of human vision is about 10^{-6} lux [390].

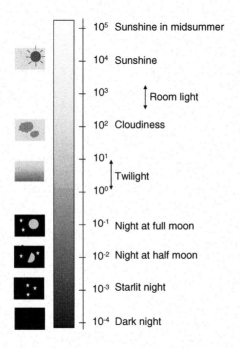

FIGURE B.1
Typical levels of illuminance for a variety of lightning conditions [6, 143].

Radiometric and photometric relation   Radiometric and photometric quantities are summarized in Table B.1 [143].

The response of a photopic eye $V(\lambda)$ is shown in Fig. B.2.

The conversion factor K from a photopic quantity to a physical quantity is ex-

TABLE B.1
Radiometric quantities vs. photometric quantities K(lm/W) [143]

Radiometric quantity	Radiometric unit	Photometric quantity	Photometric unit
Radiant intensity	W/sr*	Luminous intensity	candela (cd)
Radiant flux	W=J/S	Luminous flux	lumen (lm) =cd·sr
Irradiance	W/m²	Illuminance	lm/m² = lux
Radiance	W/m²/sr	Luminance	cd/m²

*sr:steradian

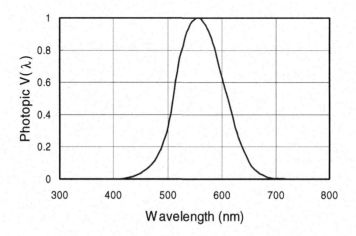

FIGURE B.2
Photopic eye response.

pressed as

$$K = 683 \frac{\int R(\lambda)V(\lambda)d\lambda}{\int R(\lambda)d\lambda}. \tag{B.1}$$

Typical conversion factors are summarized in Table B.2 [143].

Illuminance at imaging plane   Illuminance at a sensor imaging plane is described
in this section [6]. Lux is generally used as an illumination unit. It is noted that
lux is a photometric unit related to human eye characteristics, that is, it is not a pure
physical unit. Illumination is defined as the light power per unit area. Suppose an
optical system as shown in Fig. B.3. Here, a light flux F is incident on an object
whose surface is taken to be completely diffusive. When light is reflected off an
object with a perfect diffusion surface with reflectance R and area A, the reflected
light uniformly diverges at the half whole solid angle π. Thus the light flux F_o

TABLE B.2

Typical conversion factors, $K(\text{lm/W})$ [143]

Light source	Conversion factor K (lm/W)
Green 555 nm	683
Red LED	60
Daylight without clouds	140
2850 K standard light source	16
2850 K standard light source with IR filter	350

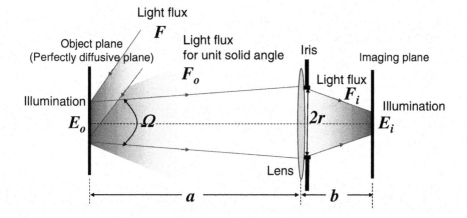

FIGURE B.3

Illumination at the imaging plane.

divergent into a unit solid angle is calculated as

$$F_o = \frac{FR}{\pi}. \tag{B.2}$$

As the solid angle to the lens aperture or iris Ω is

$$\Omega = \frac{\pi r^2}{a^2}, \tag{B.3}$$

the light flux into the sensor imaging plane through the lens with a transmittance of T, F_i, is calculated as

$$F_i = F_o \Omega = FRT \left(\frac{r}{a}\right)^2 = FRT \frac{1}{4F_N^2} \left(\frac{m}{1+m}\right)^2. \tag{B.4}$$

If the lens has a magnification factor $m = b/a$, focal length f, and F-number $F_N = f/(2r)$, then

$$a = \frac{1+m}{m} f, \tag{B.5}$$

and thus

$$\left(\frac{r}{a}\right)^2 = \left(\frac{f}{2F_N}\right)^2 \left(\frac{m}{(1+m)f}\right)^2. \tag{B.6}$$

The illuminance at the object and at the sensor imaging plane are $E_o = F/A$ and $E_i = F_i/(m^2 A)$, respectively. It is noted that the object area is focused into the sensor imaging area multiplied by the square of the magnification factor m. By using the above equations, we obtain the following relation between the illuminance at an object plane E_o and that at the sensor imaging plane E_i:

$$E_i = \frac{E_o RT}{4F_N^2(1+m)^2} \simeq \frac{E_o RT}{4F_N^2}, \tag{B.7}$$

where $m \ll 1$ is used in the second equation, which is satisfied in a conventional imaging system. For example, E_i/E_o is about 1/30 when F_N is 2.8 and $T = R = 1$. It is noted that T and R are typically less than 1, so that this ratio is typically smaller than 1/30. It is noted that the illuminance at a sensor surface decreases to $1/10 - 1/100$ of the illuminance at an object.

C

Human eye and CMOS image sensors

In this chapter, we summarize the visual processing of human eyes, because they are an ideal imaging system and a model for CMOS imaging systems. To this end, we compare the human visual system with CMOS image sensors.

Retina

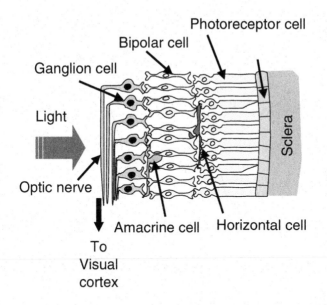

FIGURE C.1
Structure of the human retina [300].

The human eye has superior characteristics to state-of-the-art CMOS image sensors. The dynamic range of the human eye is about 200 dB with multi-resolution. In addition, the human eye has the focal plane processing function of spatio-temporal image preprocessing. In addition, humans have two eyes, which allows range finding by convergence and disparity. It is noted that distance measurements using disparity require complex processing at the visual cortex [517]. The humane retina has an area of about 5 cm × 5 cm with a thickness of 0.4 mm [187,300,392,518]. The conceptual structure is illustrated in Fig. C.1. The incident light is detected by photoreceptors, which have two types, rod and cone.

Photosensitivity of human eye

Rod photoreceptors have higher photosensitivity than cones and have adaptivity for light intensity, as shown in Fig. C.2 [519,520]. Under uniform light illumination, the rod works in a range of two orders of magnitude with saturation characteristics. Figure C.2 schematically shows a photoresponse curve under constant illumination. The photoresponse curve adaptively shifts according to the environmental illumination and eventually converts over seven orders of magnitude. The human eye has a wide dynamic range under moonlight to sunlight due to this mechanism.

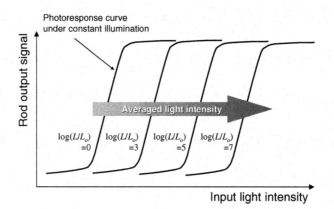

FIGURE C.2

Adaptation for light intensity in rod photoreceptors. The photoresponse curve shifts according to the average environmental illumination L. From the initial illumination L_o, the environmental illumination is changed exponentially in the range $\log(L/L_o) = 0$ to 7.

Color in retina

The human eye can sense color in a range of about 370 nm to 730 nm [392]. Rods are mainly distributed in the periphery of the retina and have a higher photosensitivity without color sensitivity, while cones are mainly concentrated in the center of the retina or fovea and have color sensitivity with less photosensitivity than rods. Thus the retina has two types of photoreceptors with high and low photosensitivities. When rods initiate vision, typically when the illumination is dark, the vision is called scotopic. When cones initiate vision, it is called photopic. The peak wavelengths for scotopic and photopic vision are 507 nm and 555 nm, respectively. For color sensitivity, rods are classified into L, M, and S types [521], which have similar characteristics of on-chip color filters in image sensors of R, G, and B, respectively. The center wavelengths of L, M, and S cones are 565 nm, 545 nm, and 440 nm, respectively [392]. The color sensitivity is different from that of animals; for example, some butterflies have a sensitivity in the ultraviolet range [390]. Surprisingly, the distribution of L, M, and S cones is not uniform [522], while image sensors have a regular arrangement of color filters, such as a Bayer pattern.

Comparison of human retina and a CMOS image sensor

Table C.1 summarizes a comparison of the human retina and a CMOS image sensor [187, 300, 392, 518].

TABLE C.1

Comparison of the human retina and a CMOS image sensor

Item	Retina	CMOS sensor
Resolution	Cones: 5×10^6 Rods: 10^8 Ganglion cells: 10^6	$1\text{--}10 \times 10^6$
Size	Rod: diameter 1 μm near fovea Cone: diameter 1–4 μm in fovea, 4–10 μm in extrafovea	2–10 μm sq.
Color	3 Cones (L, M, S) (L+M): S = 14:1	On-chip RGB filter R:G:B=1:2:1
Min. detectable illumination	$\sim$0.001 lux	0.1–1 lux
Dynamic range	Over 140 dB (adaptive)	60–70 dB
Detection method	Cis–trans isomerization $\rightarrow$two-stage amplification (500 $\times$ 500)	e–h pair generation Charge accumulation
Response time	$\sim$10 msec	Frame rate (video rate: 33 msec)
Output	Pulse frequency modulation	Analog voltage or digital
Number of outputs	Number of GCs: $\sim$1M	One analog output or bit-number in digital output
Functions	Photoelectronic conversion Adaptive function Spatio-temporal signal processing	Photoelectronic conversion Amplification Scanning

D

Fundamental characteristics of MOS capacitors

A MOS capacitor is composed of a metallic electrode (usually heavily doped poly-silicon is used) and a semiconductor with an insulator (usually SiO_2) in between. A MOS capacitor is an important part of a MOSFET and is easily implemented in standard CMOS technology by connecting the source and drain of a MOSFET. In this case, the gate and body of the MOSFET act as the electrodes of a capacitor. The characteristics of MOS capacitors are dominated by the channel underneath the insulator or SiO_2. It is noted that a MOS capacitor is a series sum of the two capacitors, a gate oxide capacitor C_{ox} and a depletion region capacitor C_D.

There are three modes in a MOS capacitor, accumulation: depletion, and inversion, as shown in Fig. D.1, which are characterized by the surface potential ψ_s [77]. The surface potential $e\psi_s$ is defined as the difference of the mid-gap energy between the surface ($z = 0$) and the bulk region ($z = \infty$).

- $\psi_s < 0$: accumulation mode

- $\psi_B > \psi_s > 0$: depletion mode

- $\psi_s > \psi_B$: inversion mode

Here $e\psi_B$ is defined as the difference between the mid-gap energy at the bulk region $E_i(\infty)$ and the Fermi energy E_{fs}. In the accumulation mode, the gate bias voltage is negative, $V_g < 0$, and holes accumulate near the surface. This mode is rarely used in image sensors. In the depletion mode, a positive gate bias $V_g > 0$ is applied and causes free carriers to be depleted near the surface region. Space charges of ionized acceptors are located in the depletion region and compensate the induced charges by the gate voltage V_g. In this mode, the surface potential ψ_s is positive but smaller than ψ_B. The third mode is the inversion mode, which is used in a MOSFET when it turns on and in CMOS image sensors to accumulate photo-generated charges. To apply a larger gate bias voltage than in the depletion mode, an inversion layer appears, where electrons accumulate in the surface region. When E_i at $z = 0$ intersects E_{fs}, the inversion mode occurs, where $\psi_s = \psi_B$. If $psi_s > 2\psi_B$, then the surface is completely inverted, that is, it becomes an n-type region in this case. This mode is called strong inversion, while the $\psi_s < 2\psi_B$ mode is called weak inversion. It is noted that the electrons in the inversion layer are thermally generated electrons and/or diffusion electrons and hence it takes some time to establish inversion layers with electrons. This means that the inversion layer in the non-equilibrium state can act as a reservoir

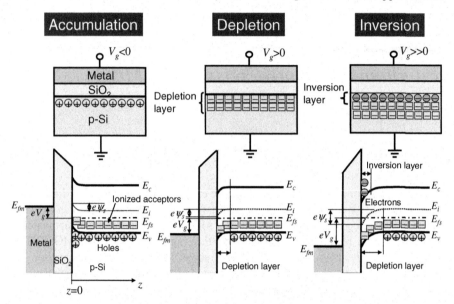

FIGURE D.1

Conceptual illustration of the three modes of MOS capacitor operation, accumulation, depletion, and inversion. E_c, E_v, E_{fs}, and E_i are the conduction band edge, valence band edge, Fermi energy of the semiconductor, and mid-gap energy, respectively. E_{fm} is the Fermi energy of the metal. V_g is the gate bias voltage.

for electrons when they are generated, for example, by incident light. This reservoir is called a potential well for photo-generated carriers. If the source and drain regions are located on either side of the inversion layer, as in a MOSFET, electrons are quickly supplied to the inversion layer from the source and drain regions, so that an inversion layer filled with electrons is established in a very short time.

E

Fundamental characteristics of MOSFET

Enhancement and depletion types

MOSFETs are classified into two types: enhancement types and depletion types. Usually in CMOS sensors, enhancement types is used, although some sensors use depletion type MOSFETs. In the enhancement type, the threshold of an NMOSFET is positive, while in the depletion type the threshold of an NMOSFET is negative. Thus depletion type NMOSFETs can turn on without an applied gate voltage, that is, a normally ON state. In a pixel circuit, the threshold voltage is critical for operation, so that some sensors use depletion type MOSFET in the pixels [116].

Operation region

The operation of MOSFETs is first classified into two regions: above threshold and below threshold (subthreshold). In each of these regions, three sub-regions exist, cutoff, linear (triode), and saturation. In the cutoff region, no drain current flows. Here we summarize the characteristics in each region for NMOSFET.

Above threshold: $V_{gs} > V_{th}$

Linear region   The condition of the linear region above threshold is

$$
\begin{aligned}
V_{gs} &> V_{th}, \\
V_{ds} &< V_{gs} - V_{th}.
\end{aligned}
\tag{E.1}
$$

In the above condition, the drain current I_d is expressed as

$$
I_d = \mu_n C_{ox} \frac{W_g}{L_g} \left[(V_{gs} - V_{th}) V_{ds} - \frac{1}{2} V_{ds}^2 \right],
\tag{E.2}
$$

where C_{ox} is the capacitance of the gate oxide per unit area and W_g and L_g are the gate width and length, respectively.

Saturation region

$$V_{gs} > V_{th},$$
$$V_{ds} > V_{gs} - V_{th}. \tag{E.3}$$

In the above condition,

$$I_d = \frac{1}{2}\mu_n C_{ox}\frac{W_g}{L_g}\left(V_{gs} - V_{th}\right)^2. \tag{E.4}$$

For short channel transistors, channel length modulation effect must be considered, and thus Eq. E.4 is modified as [126]

$$I_d = \frac{1}{2}\mu_n C_{ox}\frac{W_g}{L_g}\left(V_{gs} - V_{th}\right)^2\left(1 + \lambda V_{ds}\right)$$
$$= I_{sat}(V_{gs})\left(1 + \lambda V_{ds}\right), \tag{E.5}$$

where $I_{sat}(V_{gs})$ is the saturation drain current at V_{gs} without the channel length modulation effect. Equation E.5 means that even in the saturation region, the drain current gradually increases according to the drain–source voltage. In a bipolar transistor, a similar effect is called the Early effect and the characteristics parameter is the Early voltage V_E. In a MOSFET, the Early voltage V_E is $1/\lambda$ from Eq. E.5.

Subthreshold region

In this region, the following condition is satisfied:

$$0 < V_{gs} < V_{th}. \tag{E.6}$$

In this condition, the drain current still flows and is expressed as [523]

$$I_d = I_o \exp\left[\frac{e}{mk_BT}\left(V_{gs} - V_{th} - \frac{mk_BT}{e}\right)\right]\left[1 - \exp\left(-\frac{e}{k_BT}V_{ds}\right)\right], \tag{E.7}$$

Here m is the body-effect coefficient [524], defined later, and I_o is given by

$$I_o = \mu_n C_{ox}\frac{W_g}{L_g}\frac{1}{m}\left(\frac{mk_BT}{e}\right)^2. \tag{E.8}$$

The intuitive method to extract the above subthreshold current is given in Ref. [42, 171]. Some smart image sensors utilize the subthreshold operation, and hence we briefly consider the origin of the subthreshold current after the treatment in Refs. [42, 171].

The drain current in the subthreshold region originates from the diffusion current, which is caused by differences of electron density between the source and drain ends, n_s and n_d, that is,

$$I_d = -qW_g x_c D_n\frac{n_d - n_s}{L_g}, \tag{E.9}$$

where x_c is the channel depth. It is noted that the electron density at each end $n_{s,d}$ is determined by the electron barrier height between each end and the flat channel $\Delta E_{s,d}$, and thus

$$n_{s,d} = n_o \exp\left(-\frac{\Delta E_{s,d}}{k_B T}\right), \tag{E.10}$$

where n_o is a constant. The energy barrier in each end is given by

$$\Delta E_{s,d} = -e\psi_s + e(V_{bi} + V_{s,d}), \tag{E.11}$$

where ψ_s is the surface potential at the gate. In the subthreshold region, ψ_s is roughly a linear function of the gate voltage V_{gs} as

$$\psi_s = \psi_o + \frac{V_{gs}}{m}. \tag{E.12}$$

Here m is the body-effect coefficient given by

$$m = 1 + \frac{C_d}{C_{ox}}, \tag{E.13}$$

were C_d is the capacitance of the depletion layer per unit area. It is noted that $1/m$ is a measure of the capacitive coupling ratio from gate to channel. By using Eqs. E.9, E.10, E.11, and E.12, Eq. E.7 is obtained when the source voltage is connected to the ground voltage. The subthreshold slope S is conventionally used for the measurement of the subthreshold characteristics and defined as

$$S = \left(\frac{d(\log_{10} I_{ds})}{dV_{gs}}\right)^{-1} = 2.3\frac{mk_B T}{e} = 2.3\frac{k_B T}{e}\left(1 + \frac{C_d}{C_{ox}}\right) \tag{E.14}$$

The value of S is typically 70–100 mV/decade.

Linear region   The subthreshold region is also classified into linear and saturation regions, as well as the region above threshold. In the linear region, I_d depends on V_{ds}, while in the saturation region, I_d is almost independent of V_{ds}.

In the linear region, V_{ds} is small and both diffusion currents from the drain and source contribute to the drain current. Expanding Eq. E.7 under the condition $V_{ds} < k_B T/e$ gives

$$I_d = I_o \exp\left[\frac{e}{mk_B T}(V_{gs} - V_{th})\right]\frac{e}{k_B T}V_{ds}, \tag{E.15}$$

which shows that I_d is linear with V_{ds}.

Saturation region   In this region, V_{ds} is larger than $k_B T/e$ and thus Eq. E.7 becomes

$$I_d = I_o \exp\left[\frac{e}{mk_B T}(V_{gs} - V_{th})\right]. \tag{E.16}$$

In this region, the drain current is independent of the drain–source voltage and only depends on the gate voltage when the source voltage is constant. The transition from the linear to the saturation region occurs around $V_{ds} \approx 4k_B T/e$, which is about 100 mV at room temperature [171].

F

Optical format and resolution

TABLE F.1
Optical Format [525]

Format	Diagonal (mm)	H (mm)	V (mm)	Comment
$1/n$ inch	$16/n$	$12.8/n$	$9.6/n$	$n < 3$
$1/n$ inch	$18/n$	$14.4/n$	$10.8/n$	$n \geq 3$
35 mm	43.27	36.00	24.00	a.r.* 3:2
APS-C	27.26	22.7	15.1	
Four-thirds	21.63	17.3	13.0	a.r. 4:3

*aspect ratio

TABLE F.2
Resolution [2]

Abbreviation	Full name	Pixel number
CIF	Common Intermediate Format	352×288
QCIF	Quarter Common Intermediate Format	176×144
VGA	Video Graphics Array	640×480
QVGA	Quarter VGA	320×240
SVGA	Super VGA	800×400
XGA	eXtended Graphics Array	1024×768
UXGA	Ultra XGA	1600×1200

References

[1] A. Moini. *Vision Chips*. Kluwer Acadmic Publisher, Dordrecht, The Netherlands, 2000.

[2] J. Nakamura, editor. *Image Sensors and Signal Processing for Digital Still Cameras*. CRC Press, Boca Raton, FL, 2005.

[3] K. Yonemoto. *Fundamentals and Applications of CCD/CMOS Image Sensors*. CQ Pub. Co., Ltd., Tokyo, Japan, 2003. In Japanese.

[4] O. Yadid-Pecht and R. Etinne-Cummings, editors. *CMOS Imagers: From Phototransduction to Image Processing*. Kluwer Academic Publishers, Dordrecht, The Netherlands, 2004.

[5] Y. Takemura. *CCD Camera Technologies*. Corona Pub., Co. Ltd., Tokyo, Japan, 1997. In Japanese.

[6] Y. Kiuchi. *Fundamentals and Applications of Image Sensors*. The Nikkankogyo Shimbun Ltd., Tokyo, Japan, 1991. In Japanese.

[7] T. Ando and H. Komobuchi. *Introduction of Solid-State Image Sensors*. Nihon Riko Syuppan Kai, Tokyo, Japan, 1999. In Japanese.

[8] A.J.P. Theuwissen. *Solid-State Imaging with Charge-Coupled Devices*. Kluwer Academic Pub., Dordrecht, The Netherlands, 1995.

[9] S. Morrison. A new type of photosensitive junction device. *Solid-State Electron.*, 5:485–494, 1963.

[10] J.W. Horton, R.V. Mazza, and H. Dym. The scanistor — A solid-state image scanner. *Proc. IEEE*, 52(12):1513–1528, December 1964.

[11] P.K. Weimer, G. Sadasiv, J.E. Meyer, Jr., L. Meray-Horvath, and W.S. Pike. A self-scanned solid-state image sensor. *Proc. IEEE*, 55(9):1591–1602, September 1967.

[12] M.A. Schuster and G. Strull. A monolithic mosaic of photon sensors for solid-state imaging applications. *IEEE Trans. Electron Devices*, ED-13(12):907–912, December 1966.

[13] G.P. Weckler. Operation of p-n Junction Photodetectors in a Photon Flux Integration Mode. *IEEE J. Solid-State Circuits*, SC-2(3):65–73, September 1967.

[14] R.H. Dyck and G.P. Weckler. Integrated arrays of silicon photodetectors for image sensing. *IEEE Trans. Electron Devices*, ED-15(4):196–201, April 1968.

[15] P.J. Noble. Self-scanned silicon image detector arrays. *IEEE Trans. Electron Devices*, ED-15(4):202–209, April 1968.

[16] R.A. Anders, D.E. Callahan, W.F. List, D.H. McCann, and M.A. Schuster. Developmental solid-state imaging system. *IEEE Trans. Electron Devices*, ED-15(4):191–196, April 1968.

[17] G. Sadasiv, P.K. Weimer, and W.S. Pike. Thin-film circuits for scanning image-sensor arrays. *IEEE Trans. Electron Devices*, ED-15(4):215–219, April 1968.

[18] W.F. List. Solid-state imaging — Methods of approach. *IEEE Trans. Electron Devices*, ED-15(4):256–261, April 1968.

[19] W.S. Boyle and G.E. Smith. Charge-Coupled Semiconductor Devices. *Bell System Tech. J.*, 49:587–593, 1970.

[20] G.F. Amelio, M.F. Tompsett, and G.E. Smith. Experimental Verification of the Charge-Coupled Semiconductor Device Concept. *Bell System Tech. J.*, 49:593–600, 1970.

[21] G.E. Smith. The invention of the CCD. *Nuclear Instruments & Methods in Physics Research A*, 471:1–5, 2001.

[22] N. Koike, I. Takemoto, K. Satoh, S. Hanamura, S. Nagahara, and M. Kubo. MOS Area Sensor: Part I — Design Consideration and Performance of an n-p-n Structure 484 × 384 Element Color MOS Imager. *IEEE Trans. Electron Devices*, 27(8):1682–1687, August 1980.

[23] H. Nabeyama, S. Nagahara, H. Shimizu, M. Noda, and M. Masuda. All Solid State Color Camera with Single-Chip MOS Imager. *IEEE Trans. Consumer Electron.*, CE-27(1):40–46, February 1981.

[24] S. Ohba, M. Nakai, H. Ando, S. Hanamura, S. Shimada, K. Satoh, K. Takahashi, M. Kubo, and T. Fujita. MOS Area Sensor: Part II — Low-Noise MOS Area Sensor with Antiblooming Photodiodes. *IEEE J. Solid-State Circuits*, 15(4):747–752, August 1980.

[25] M. Aoki, H. Ando, S. Ohba, I. Takemoto, S. Nagahara, T. Nakano, M. Kubo, and T. Fujita. 2/3-Inch Format MOS Single-Chip Color Imager. *IEEE J. Solid-State Circuits*, 17(2):375–380, April 1982.

[26] H. Ando, S. Ohba, M. Nakai, T. Ozaki, N. Ozawa, K. Ikeda, T. Masuhara, T. Imaide, I. Takemoto, T. Suzuki, and T. Fujita. Design consideration and performance of a new MOS imaging device. *IEEE Trans. Electron Devices*, ED-32(8):1484–1489, August 1985.

[27] T. Kinugasa, M. Noda, T. Imaide, I. Aizawa, Y. Todaka, and M Ozawa. An Electronic Variable-Shutter System in Video Camera Use. *IEEE Trans. Consumer Electron.*, CE-33(3):249–255, August 1987.

[28] K. Senda, S. Terakawa, Y. Hiroshima, and T. Kunii. Analysis of charge-priming transfer efficiency in CPD image sensors. *IEEE Trans. Electron Devices*, 31(9):1324–1328, September 1984.

[29] T. Nakamura, K. Matsumoto, R. Hyuga, and A Yusa. A new MOS image sensor operating in a non-destructive readout mode. In *Tech. Dig. Int'l Electron Devices Meeting (IEDM)*, pages 353–356, 1986.

[30] J. Hynecek. A new device architecture suitable for high-resolution and high-performance image sensors. *IEEE Trans. Electron Devices*, 35(5):646–652, May 1988.

[31] N. Tanaka, T. Ohmi, and Y. Nakamura. A novel bipolar imaging device with self-noise-reduction capability. *IEEE Trans. Electron Devices*, 36(1):31–38, January 1989.

[32] A. Yusa, J. Nishizawa, M. Imai, H. Yamada, J. Nakamura, T. Mizoguchi, Y. Ohta, and M. Takayama. SIT Image sensor: Design considerations and characteristics. *IEEE Trans. Electron Devices*, 33(6):735–742, June 1986.

[33] F. Andoh, K. Taketoshi, K. Nakamura, and M. Imai. Amplified MOS Intelligent Imager. *J. Inst. Image Information & Television Eng.*, 41(11):1075–1082, November 1987. In Japanese.

[34] F. Andoh, K. Taketoshi, J. Yamazaki, M. Sugawara, Y. Fujita, K. Mitani, Y. Matuzawa, K. Miyata, and S. Araki. A 250000-pixel image sensor with FET amplification at each pixel for high-speed television cameras. *Dig. Tech. Papers Int'l Solid-State Circuits Conf. (ISSCC)*, pages 212–213, 298, February 1990.

[35] M. Kyomasu. A New MOS Imager Using Photodiode as Current Source. *IEEE J. Solid-State Circuits*, 26(8):1116–1122, August 1991.

[36] M. Sugawara, H. Kawashima, F. Andoh, N. Murata, Y. Fujita, and M. Yamawaki. An amplified MOS imager suited for image processing. *Dig. Tech. Papers Int'l Solid-State Circuits Conf. (ISSCC)*, pages 228–229, February 1994.

[37] M. Yamawaki, H. Kawashima, N. Murata, F. Andoh, M. Sugawara, and Y. Fujita. A pixel size shrinkage of amplified MOS imager with two-line mixing. *IEEE Trans. Electron Devices*, 1996.

[38] E. R. Fossum. Active pixel sensors — Are CCD's dinosaurs? In *Proc. SPIE*, volume 1900 of *Charge-Coupled Devices and Optical Sensors III*, pages 2–14, 1993.

[39] E.R. Fossum. CMOS Image Sensors: Electronic Camera-On-A-Chip. *IEEE Trans. Electron Devices*, 44(10):1689–1698, October 1997.

[40] R.M. Guidash, T.-H. Lee, P.P.K. Lee, D.H. Sackett, C.I. Drowley, M.S. Swenson, L. Arbaugh, R. Hollstein, F. Shapiro, and S. Domer. A 0.6 μm CMOS pinned photodiode color imager technology. In *Tech. Dig. Int'l Electron Devices Meeting (IEDM)*, pages 927–929, December 1997.

[41] H.S. Wong. Technology and device scaling considerations for CMOS imagers. *IEEE Trans. Electron Devices*, 43(12):2131–2141, December 1996.

[42] C. Mead. *Analog VLSI and Neural Systems*. Addison-Wesley Publishing Company, Reading, MA, 1989.

[43] C. Koch and H. Liu. *VISION CHIPS, Implementing Vision Algorithms with Analog VLSI Circuits*. IEEE Computer Society, Los Alamitos, CA, 1995.

[44] T.M. Bernard, B.Y. Zavidovique, and F.J. Devos. A Programmbale Artificial Retina. *IEEE J. Solid-State Circuits*, 28(7):789–798, July 1993.

[45] S. Kameda and T. Yagi. An analog VLSI chip emulating sustained and transient response channels of the vertebrate retina. *IEEE Trans. Neural Networks*, 14(5):1405–1412, September 2003.

[46] R. Takami, K. Shimonomura, S. Kamedaa, and T. Yagi. An image preprocessing system employing neuromorphic 100 × 100 pixel silicon retina. In *Int'l Symp. Circuits & Systems (ISCAS)*, pages 2771–2774, Kobe, Japan, May 2005.

[47] www.itrs.net/.

[48] G. Moore. Cramming more components onto integrated circuits. *Electronics*, 38(8), April 1965.

[49] L. Fortuna, P. Arena, D. Balya, and A. Zarandy. Cellular neural networks: a paradigm for nonlinear spatio-temporal processing. *IEEE Circuits & Systems Mag.*, 1(4):6–21, 2001.

[50] S. Espejo, A. Rodríguez-Vázquez, R. Domínguez-Castro, J.L. Huertas, and E. Sánchez-Sinencio. Smart-pixel cellular neural networks in analog current-mode CMOS technology. *IEEE J. Solid-State Circuits*, 29(8):895–905, August 1994.

[51] R. Domínguez-Castro, S. Espejo, A. Rodríguez-Vázquez, R.A. Carmona, P. Földesy, A. Zárandy, P. Szolgay, T. Szirányi, and T. Roska. A 0.8-μm C-MOS two-dimensional programmable mixed-signal focal-plane array processor with on-chip binary imaging and instructions storage. *IEEE J. Solid-State Circuits*, 32(7):1013–1026, July 1997.

[52] G. Liñan-Cembrano, L. Canranza, S. Espejo, R. Dominguez-Castro, and A. Rodíguez-Vázquez. CMOS mixed-signal flexible vision chips. In M. Valle,

editor, *Smart Adaptive Systems on Silicon*, pages 103–118. Kluwer Academic Pub., Dordecht, The Netherlands, 2004.

[53] R. Etienne-Cummings, Z.K. Kalayjian, and D. Cai. A Programmable Focal-Plane MIMD Image Processor Chip. *IEEE J. Solid-State Circuits*, 36(1):64–73, January 2001.

[54] E. Culurciello, R. Etienne-Cummings, and K.A. Boahen. A Biomorphic Digital Image Sensor. *IEEE J. Solid-State Circuits*, 38(2):281–294, February 2003.

[55] P. Dudek and P.J. Hicks. A General-Purpose Processor-per-Pixel Analog SIMD Vision Chip. *IEEE Trans. Circuits & Systems I*, 52(1):13–20, January 2005.

[56] P. Dudek and P.J. Hicks. A CMOS General-Purpose Sampled-Data Analog Processing Element. *IEEE Trans. Circuits & Systems II*, 47(5):467–473, May 2000.

[57] M. Ishikawa, K. Ogawa, T. Komuro, and I. Ishii. A CMOS vision chip with SIMD processing element array for 1 ms image processing. In *Dig. Tech. Papers Int'l Solid-State Circuits Conf. (ISSCC)*, pages 206–207, February 1999.

[58] M. Ishikawa and T. Komuro. Digital vision chips and high-speed vision systems. In *Dig. Tech. Papers Symp. VLSI Circuits*, pages 1–4, June 2001.

[59] N. Mukohzaka, H. Toyoda, S. Mizuno, M.H. Wu, Y. Nakabo, and M. Ishikawa. Column parallel vision system: CPV. In *Proc. SPIE*, volume 4669, pages 21–28, San Jose, CA, January 2002.

[60] T. Komuro, S. Kagami, and M. Ishikawa. A high speed digital vision chip with multi-grained parallel processing capability. In *IEEE Workshop on Charge-Coupled Devices & Advanced Image Sensors*, Elmau, Geramany, June 2003.

[61] T. Komuro, I. Ishii, M. Ishikawa, and A. Yoshida. A Digital Vision Chip Specialized for High-Speed Target Tracking. *IEEE Trans. Electron Devices*, 50(1):191–199, January 2003.

[62] T. Komuro, S. Kagami, and M. Ishikawa. A dynamically reconfigurable SIMD processor for a vision chip. *IEEE J. Solid-State Circuits*, 39(1):265–268, January 2004.

[63] T. Komuro, S. Kagami, M. Ishikawa, and Y. Katayama. Development of a Bit-level Compiler for Massively Parallel Vision Chips. In *IEEE Int'l Workshop Computer Architecture for Machine Perception (CAMP)*, pages 204–209, Palermo, July 2005.

[64] D. Renshaw, P.B. Denyer, G. Wang, and M. Lu. ASIC VISION. In *Proc. Custom Integrated Circuits Conf. (CICC)*, pages 7.3/1–7.3/4, May 1990.

[65] P.B. Denyer, D. Renshaw, Wang G., and Lu M. CMOS image sensors for multimedia applications. In *Proc. Custom Integrated Circuits Conf. (CICC)*, pages 11.5.1–11.5.4, May 1993.

[66] K. Chen, M. Afghani, P.E. Danielsson, and C. Svensson. PASIC: A processor-A/D converter-sensor integrated circuit. In *Int'l Symp. Circuits & Systems (ISCAS)*, volume 3, pages 1705–1708, May 1990.

[67] J.-E. Eklund, C. Svensson, and A. Åström. VLSI Implementation of a Focal Plane Image Processor – A Realization of the Near-Sensor Image Processing Concept. *IEEE Trans. VLSI Systems*, 4(3):322–335, September 1996.

[68] L. Lindgren, J. Melander, R. Johansson, and B. Moller. A multiresolution 100-GOPS 4-Gpixels/s programmable smart vision sensor for multisense imaging. *IEEE J. Solid-State Circuits*, 40(6):1350–1359, June 2005.

[69] E.D. Palik. *Handbook of Optical Constants of Solids*. Academic Press, New York, NY, 1977.

[70] S.E. Swirhun, H.-H. Kwark, and R.M. Swanson. Measurement of electron lifetime, electron mobility and band-gap narrowing in heavily doped p-type silicon. In *Tech. Dig. Int'l Electron Devices Meeting (IEDM)*, pages 24–27, 1986.

[71] J. del Alamo, S. Swirhun, and R.M. Swanson. Measuring and modeling minority carrier transport in heavily doped silicon. *Solid-State Electronic.*, 28:47–54, 1985.

[72] J. del Alamo, S. Swirhun, and R.M. Swanson. Simultaneous measurement of hole lifetime, hole mobility and bandgap narrowing in heavily doped n-type silicon. In *Tech. Dig. Int'l Electron Devices Meeting (IEDM)*, pages 290–293, 1985.

[73] J.S. Lee, R.I. Hornsey, and D. Renshaw. Analysis of CMOS Photodiodes. I. Quantum efficiency. *IEEE Trans. Electron Devices*, 50(5):1233–1238, May 2003.

[74] J.S. Lee, R.I. Hornsey, and D. Renshaw. Analysis of CMOS Photodiodes. II. Lateral photoresponse. *IEEE Trans. Electron Devices*, 50(5):1239–1245, May 2003.

[75] S.G. Chamberlain, D.J. Roulston, and S.P. Desai. Spectral Response Limitation Mechanisms of a Shallow Junction n+-p Photodiode. *IEEE J. Solid-State Circuits*, SC-13(1):167–172, February 1978.

[76] I. Murakami, T. Nakano, K. Hatano, Y. Nakashiba, M. Furumiya, T. Nagata, H. Utsumi, S. Uchida, K. Arai, N. Mutoh, A. Kohno, N. Teranishi, and Y. Hokari. New Technologies of Photo-Sensitivity Improvement and VOD Shutter Voltage Reduction for CCD Image Sensors. In *Proc. SPIE*, volume 3649, pages 14–21, San Jose, CA, 1999.

[77] S.M. Sze. *Physics of Semiconductor Devices*. John Wiley & Sons, Inc., New York, NY, 1981.

[78] N.V. Loukianova, H.O. Folkerts, J.P.V. Maas, D.W.E. Verbugt, A.J. Mierop, W. Hoekstra, E. Roks, and A.J.P. Theuwissen. Leakage current modeling of test structures for characterization of dark current in CMOS image sensors. *IEEE Trans. Electron Devices*, 50(1):77–83, January 2003.

[79] B. Pain, T. Cunningham, B. Hancock, C. Wrigley, and C. Sun. Excess Noise and Dark Current Mechanism in CMOS Imagers. In *IEEE Workshop on Charge-Coupled Devices & Advanced Image Sensors*, pages 145–148, Karuizawa, Japan, June 2005.

[80] A.S. Grove. *Physics and Technology of Semiconductor Devices*. John Wiley & Sons, Inc., New York, NY, 1967.

[81] G.A.M. Hurkx, H.C. de Graaff, W.J. Kloosterman, and M.P.G. Knuvers. A new analytical diode model including tunneling and avalanche breakdown. *IEEE Trans. Electron Devices*, 39(9):2090–2098, September 1992.

[82] H.O. Folkerts, J.P.V. Maas, D.W.E. Vergugt, A.J. Mierop, W. Hoekstra, N.V. Loukianova, and E. Rocks. Characterization of Dark Current in CMOS Image Sensors. In *IEEE Workshop on Charge-Coupled Devices & Advanced Image Sensors*, Elmau, Germany, May 2003.

[83] H. Zimmermann. *Silicon Optoelectronic Integrated Cicuits*. Springer-Verlag, Berlin, Germany, 2004.

[84] S. Radovanović, A.-J. Annema, and B. Nauta. *High-Speed Photodiodes in Standard CMOS Technology*. Springer, Dordecht, The Netherlands, 2006.

[85] B. Razavi. *Design of Integrated Circuits for Optical Communications*. McGraw-Hill Companies Inc., New York, NY, 2003.

[86] J. Singh. *Semiconductor Optoelectronics: Physics and Technology*. McGraw-Hill, Inc., New York, NY, 1995.

[87] V. Brajovic, K. Mori, and N. Jankovic. 100frames/s CMOS Range Image Sensor. In *Dig. Tech. Papers Int'l Solid-State Circuits Conf. (ISSCC)*, pages 256–257, February 2001.

[88] Y. Takiguchi, H. Maruyama, M. Kosugi, F. Andoh, T. Kato, K. Tanioka, J. Yamazaki, K. Tsuji, and T. Kawamura. A CMOS Imager Hybridized to an Avalanche Multiplied Film. *IEEE Trans. Electron Devices*, 44(10):1783–1788, October 1997.

[89] A. Biber, P. Seitz, and H. Jäckel. Avalanche Photodiode Image Sensor in Standard BiCMOS Technology. *IEEE Trans. Electron Devices*, 47(11):2241–2243, November 2000.

[90] J.C. Jackson, P.K. Hurley, A.P. Morrison, B. Lane, and A. Mathewson. Comparing Leakage Currents and Dark Count Rates in Shallow Junction Geiger-Mode Avalanche Photodiodes. *Appl. Phys. Lett.*, 80(22):4100–4102, June 2002.

[91] J. C. Jackson, A. P. Morrison, D. Phelan, and A. Mathewson. A Novel Silicon Geiger-Mode Avalanche Photodiode. In *Tech. Dig. Int'l Electron Devices Meeting (IEDM)*, December 2002.

[92] J. C. Jackson, D. Phelan, A. P. Morrison, M. Redfern, and A. Mathewson. Towards integrated single photon counting arrays. *Opt. Eng.*, 42(1):112–118, January 2003.

[93] A.M. Moloney, A.P. Morrison, J.C. Jackson, A. Mathewson, J. Alderman, J. Donnelly, B. O'Neill, A.-.M. Kelleher, G. Healy, and P.J. Murphy. Monolithically Integrated Avalanche Photodiode and Transimpedance Amplifier in a Hybrid Bulk/SOI CMOS Process. *Electron. Lett.*, 39(4):391–392, February 2003.

[94] J. C. Jackson, J. Donnelly, B. O'Neill, A-M. Kelleher, G. Healy, A. P. Morrison, and A. Mathewson. Integrated Bulk/SOI APD Sensor: Bulk Substrate Inspection with Geiger-Mode Avalanche Photodiodes. *Electron. Lett.*, 39(9):735–736, May 2003.

[95] A. Rochas, A.R. Pauchard, P.-A. Besse, D. Pantic, Z. Prijic, and R.S. Popovic. Low-Noise Silicon Avalanche Photodiodes Fabricated in Conventional CMOS Technologies. *IEEE Trans. Electron Devices*, 49(3):387–394, March 2002.

[96] A. Rochas, M. Gosch, A. Serov, P.A. Besse, R.S. Popovic, T. Lasser, and R. Rigler. First Fully Integrated 2-D Array of Single-Photon Detectors in Standard CMOS Technology. *IEEE Photon. Tech. Lett.*, 15(7):963–965, July 2003.

[97] C. Niclass, A. Rochas, P.-A. Besse, and E. Charbon. Toward a 3-D Camera Based on Single Photon Avalanche Diodes. *IEEE Selcted Topic Quantum Electron.*, 2004.

[98] C. Niclass, A. Rochas, P.-A. Besse, and E. Charbon. Design and Characterization of a CMOS 3-D Image Sensor Based on Single Photon Avalanche Diodes. *IEEE J. Solid-State Circuits*, 40(9):1847–1854, September 2005.

[99] S. Bellis, R. Wilcock, and C. Jackson. Photon Counting Imaging: the DigitalAPD. In *SPIE-IS&T Electronic Imaging: Sensors, Cameras, and Systems for Scientific/Industrial Applications VII*, volume 6068, pages 60680D–1–D–10, San Jose, CA, January 2006.

[100] C.J. Stapels, W.G. Lawrence, F.L. Augustine, and J.F. Christian. Characterization of a CMOS Geiger Photodiode Pixel. *IEEE Trans. Electron Devices*, 53(4):631–635, April 2006.

[101] H. Finkelstein, M.J. Hsu, and S.C. Esener. STI-Bounded Single-Photon Avalanche Diode in a Deep-Submicrometer CMOS Technology. *IEEE Electron Device Lett.*, 27(11):887–889, November 2006.

[102] M. Kubota, T. Kato, S. Suzuki, H. Maruyama, K. Shidara, K. Tanioka, K. Sameshima, T. Makishima, K. Tsuji, , T. Hirai, and T. Yoshida. Ultra high-sensitivity new super-HARP camera. *IEEE Trans. Broadcast.*, 42(3):251–258, September 1996.

[103] T. Watabe, M. Goto, H. Ohtake, H. Maruyama, M. Abe, K. Tanioka, and N. Egami. New signal readout method for ultra high-sensitivity CMOS image sensor. *IEEE Trans. Electron Devices*, 50(1):63–69, January 2003.

[104] S. Aihara, Y. Hirano, T. Tajima, K. Tanioka, M. Abe, N. Saito, N. Kamata, and D. Terunuma. Wavelength selectivities of organic photoconductive films: Dye-doped polysilanes and zinc phthalocyanine/tris-8-hydroxyquinoline aluminum double layer. *Appl. Phys. Lett.*, 82(4):511–513, January 2003.

[105] T. Watanabe, S. Aihara, N. Egami, M. Kubota, K. Tanioka, N. Kamata, and D. Terunuma. CMOS Image Sensor Overlaid with an Organic Photoconductive Film. In *IEEE Workshop on Charge-Coupled Devices & Advanced Image Sensors*, pages 48–51, Karuizawa, Japan, June 2005.

[106] S. Takada, M. Ihama, and M. Inuiya. CMOS Image Sensor with Organic Photoconductive Layer Having Narrow Absorption Band and Proposal of S-tack Type Solid-State Image Sensors. In *Proc. SPIE*, volume 6068, pages 60680A–1–A8, San Jose, CA, January 2006.

[107] J. Burm, K.I. Litvin, D.W. Woodard, W.J. Schaff, P. Mandeville, M.A. Jaspan, M.M. Gitin, and L.F. Eastman. High-frequency, high-efficiency MSM photodetectors. *IEEE J. Quantum Electron.*, 31(8):1504–1509, August 1995.

[108] K. Tanaka, F. Ando, K. Taketoshi, I. Ohishi, and G. Asari. Novel Digital Photosensor Cell in GaAs IC Using Conversion of Light Intensity to Pulse Frequency. *Jpn. J. Appl. Phys.*, 32(11A):5002–5007, November 1993.

[109] E. Lange, E. Funatsu, J. Ohta, and K. Kyuma. Direct image processing using arrays of variable-sensitivity photodetectors. In *Dig. Tech. Papers Int'l Solid-State Circuits Conf. (ISSCC)*, pages 228–229, February 1995.

[110] H.-B. Lee, H.-S. An, H.-I. Cho, J.-H. Lee, and S.-H. Hahm. UV photo-responsive characteristics of an n-channel GaN Schottky-barrier MISFET for UV image sensors. *IEEE Electron Device Lett.*, 27(8):656–658, August 2006.

[111] M. Abe. Image sensors and circuit technologies. In T. Enomoto, editor, *Video/Image LSI System Design Technology*, pages 208–248. Corona Pub., Co. Ltd., Tokyo, Japan, 2003. In Japanese.

[112] N. Teranishi, A. Kohono, Y. Ishihara, E. Oda, and K. Arai. No image lag photodiode structure in the interline CCD image sensor. In *Tech. Dig. Int'l Electron Devices Meeting (IEDM)*, pages 324–327, 1982.

[113] B.C. Burkey, W.C. Chang, J. Littlehale, T.H. Lee, T.J. Tredwell, J.P. Lavine, and E.A. Trabka. The pinned photodiode for an interline-transfer CCD image sensor. In *Tech. Dig. Int'l Electron Devices Meeting (IEDM)*, pages 28–31, 1984.

[114] I. Inoue, H. Ihara, H. Yamashita, T. Yamaguchi, H. Nozaki, and R. Miyagawa. Low dark current pinned photo-diode for CMOS image sensor. In *IEEE Workshop on Charge-Coupled Devices & Advanced Image Sensors*, pages 25–32, Karuizawa, Japan, June 1999.

[115] S. Agwani, R Cichomski, M. Gorder, A. Niederkorn, Sknow M., and K. Wanda. A 1/3" VGA CMOS Imaging System On a Chip. In *IEEE Workshop on Charge-Coupled Devices & Advanced Image Sensors*, pages 21–24, Karuizawa, Japan, June 1999.

[116] K. Yonemoto and H. Sumi. A CMOS image sensor with a simple fixed-pattern-noise-reduction technology and a hole accumulation diode. *IEEE J. Solid-State Circuits*, 2000.

[117] M. Noda, T. Imaide, T. Kinugasa, and R. Nishimura. A Solid State Color Video Camera with a Horizontal Readout MOS Imager. *IEEE Trans. Consumer Electron.*, CE-32(3):329–336, August 1986.

[118] S. Miyatake, M. Miyamoto, K. Ishida, T. Morimoto, Y. Masaki, and H. Tanabe. Transversal-readout architecture for CMOS active pixel image sensors. *IEEE Trans. Electron Devices*, 50(1):121–129, January 2003.

[119] J.D. Plummer and J.D. Meindl. MOS electronics for a portable reading aid for the blind. *IEEE J. Solid-State Circuits*, 7(2):111–119, April 1972.

[120] L.J. Kozlowski, J. Luo, W.E. Kleinhans, and T. Liu. Comparison of Passive and Active Pixel Schemes for CMOS Visible Imagers. In *Proc. SPIE*, volume 3360 of *Infrared Readout Electronics IV*, pages 101–110, Orland, FL, April 1998.

[121] Y. Endo, Y. Nitta, H. Kubo, T. Murao, K. Shimomura, M. Kimura, K. Watanabe, and S. Komori. 4-micron pixel CMOS iamge sensor with low image lag and high-temperature operability. In *Proc. SPIE*, volume 5017, pages 196–204, Santa Clara, CA, January 2003.

[122] I. Inoue, N. Tanaka, H. Yamashita, T. Yamaguchi, H. Ishiwata, and H. Ihara. Low-leakage-current and low-operating-voltage buried photodiode for a C-MOS imager. *IEEE Trans. Electron Devices*, 50(1):43–47, January 2003.

[123] O. Yadid-Pecht, B. Pain, C. Staller, C. Clark, and E. Fossum. CMOS active pixel sensor star tracker with regional electronic shutter. *IEEE J. Solid-State Circuits*, 32(2):285–288, February 1997.

[124] S.E. Kemeny, R. Panicacci, B. Pain, L. Matthies, and E.R. Fossum. Multiresolution Image Sensor. *IEEE Trans. Circuits & Systems Video Tech.*, 7(4):575–583, August 1997.

[125] K. Salama and A. El Gamal. Analysis of active pixel sensor readout circuit. *IEEE Trans. Circuits & Systems I*, 50(7):941–945, July 2003.

[126] B. Razavi. *Design of Analog CMOS Integrated Circuits*. McGraw-Hill Companies, Inc., New York, NY, 2001.

[127] J. Hynecek. Analysis of the photosite reset in FGA image sensors. *IEEE Trans. Electron Devices*, 37(10):2193–2200, October 1990.

[128] S. Mendis, S.E. Kemeny, and E.R. Fossum. CMOS active pixel image sensor. *IEEE Trans. Electron Devices*, 41(3):452–453, March 1994.

[129] R.H. Nixon, S.E. Kemeny, B. Pain, C.O. Staller, and E.R. Fossum. 256 × 256 CMOS active pixel sensor camera-on-a-chip. *IEEE J. Solid-State Circuits*, 31(12):2046–2050, December 1996.

[130] T. Sugiki, S. Ohsawa, H. Miura, M. Sasaki, N. Nakamura, I. Inoue, M. Hoshino, Y. Tomizawa, and T. Arakawa. A 60 mW 10 b CMOS image sensor with column-to-column FPN reduction. In *Dig. Tech. Papers Int'l Solid-State Circuits Conf. (ISSCC)*, pages 108–109, February 2000.

[131] M.F. Snoeij, A.J.P. Theuwissen, K.A.A. Makinwa, and J.H. Huijsing. A CMOS Imager with Column-Level ADC Using Dynamic Column Fixed-Pattern Noise Reduction. *IEEE J. Solid-State Circuits*, 41(12):3007–3015, December 2006.

[132] M. Furuta, S. Kawahito, T. Inoue, and Y. Nishikawa. A cyclic A/D converter with pixel noise and column-wise offset canceling for CMOS image sensors. In *Proc. European Solid-State Circuits Conf. (ESSCIRC)*, pages 411–414, Grenoble, France, September 2005.

[133] M. Waeny, S. Tanner, S. Lauxtermann, N. Blanc, M. Willemin, M. Rechsteiner, E. Doering, J. Grupp, P. Seitz, F. Pellandini, and M. Ansorge. High sensitivity and high dynamic, digital CMOS imager. In *Proc. SPIE*, volume 406, pages 78–84, May 2001.

[134] S. Smith, J. Hurwitz, M. Torrie, D. Baxter, A. Holmes, M. Panaghiston, R. Henderson, A. Murray, S. Anderson, and P. Denyer. A single-chip 306 × 244-pixel CMOS NTSC video camera. In *Dig. Tech. Papers Int'l Solid-State Circuits Conf. (ISSCC)*, pages 170–171, February 1998.

[135] M.J. Loinaz, K.J. Singh, A.J. Blanksby, D.A. Inglis, K. Azadet, and B.D. Ackland. A 200-mW, 3.3-V, CMOS color camera IC producing 352×288 24-b video at 30 frames/s. *IEEE J. Solid-State Circuits*, 33(12):2092–2103, December 1998.

[136] Z. Zhou, B. Pain, and E.E. Fossum. CMOS active pixel sensor with on-chip successive approximation analog-to-digital converter. *IEEE Trans. Electron Devices*, 44(10):1759–1763, October 1997.

[137] I. Takayanagi, M. Shirakawa, K. Mitani, M. Sugawara, S. Iversen, J. Moholt, J. Nakamura, and E.R. Fossum. 1 1/4 inch 8.3M pixel digital output CMOS APS for UDTV application. In *Dig. Tech. Papers Int'l Solid-State Circuits Conf. (ISSCC)*, pages 216–217, February 2003.

[138] K. Findlater, R.Henderson, D. Baxter, J.E.D. Hurwitz, L. Grant, Y. Cazaux, F. Roy, D. Herault, and Y. Marcellier. SXGA pinned photodiode CMOS image sensor in 0.35 *mu*m technology. In *Dig. Tech. Papers Int'l Solid-State Circuits Conf. (ISSCC)*, page 218, February 2003.

[139] S. Decker, D. McGrath, K. Brehmer, and C.G. Sodini. A 256×256 CMOS imaging array with wide dynamic range pixels and column-parallel digital output. *IEEE J. Solid-State Circuits*, 33(12):2081–2091, December 1998.

[140] D. Yang, B. Fowler, and A. El Gamal. A Nyquist Rate Pixel Level ADC for CMOS Image Sensors. *IEEE J. Solid-State Circuits*, 34(3):348–356, March 1999.

[141] F. Andoh, H. Shimamoto, and Y. Fujita. A Digital Pixel Image Sensor for Real-Time Readout. *IEEE Trans. Electron Devices*, 47(11):2123–2127, November 2000.

[142] M. Willemin, N. Blanc, G.K. Lang, S. Lauxtermann, P. Schwider, P. Seitz, and M. Wäny. Optical characterization methods for solid-state image sensors. *Optics and Lasers Eng.*, 36(2):185–194, 2001.

[143] G.R. Hopkinson, T.M. Goodman, and S.R. Prince. *A guide to the use and calibration of detector arrya equipment*. SPIE Press, Bellingham, Washington, 2004.

[144] M.F. Snoeij, A. Theuwissen, K. Makinwa, and J.H. Huijsing. A CMOS Imager with Column-Level ADC Using Dynamic Column FPN Reduction. In *Dig. Tech. Papers Int'l Solid-State Circuits Conf. (ISSCC)*, pages 2014–2023, February 2006.

[145] J.E. Carnes and W.F. Kosonocky. Noise source in charge-coupled devices. *RCA Review*, 33(2):327–343, June 1972.

[146] B. Pain, G. Yang, M. Ortiz, C. Wrigley, B. Hancock, and T. Cunningham. Analysis and enhancement of low-light-level performance of photodiode-type CMOS active pixel imagers operated with sub-threshold reset. In *IEEE Workshop on Charge-Coupled Devices & Advanced Image Sensors*, pages 140–143, Karuizawa, Japan, June 1999.

[147] B. Pain, G. Yang, T.J. Cunningham, C. Wrigley, and B. Hancock. An Enhanced-Performance CMOS Imager with a Flushed-Reset Photodiode Pixel. *IEEE Trans. Electron Devices*, 50(1):48–56, January 2003.

[148] B.E. Bayer. Color imaging array. US patent 3,971,065, July 1976.

[149] M. Kasano, Y. Inaba, M. Mori, S. Kasuga, T. Murata, and T. Yamaguchi. A 2.0-μm Pixel Pitch MOS Image Sensor with 1.5 Transistor/Pixel and an Amorphous Si Color Filter. *IEEE Trans. Electron Devices*, 53(4):611–617, April 2006.

[150] R.D. McGrath, H. Fujita, R.M. Guidash, T.J. Kenney, and W. Xu. Shared pixels for CMOS image sensor arrays. In *IEEE Workshop on Charge-Coupled Devices & Advanced Image Sensors*, pages 9–12, Karuizawa, Japan, June 2005.

[151] H. Takahashi, M. Kinoshita, K. Morita, T. Shirai, T. Sato, T. Kimura, H. Yuzurihara, and S. Inoue. A 3.9 μm pixel pitch VGA format 10 b digital image sensor with 1.5-transistor/pixel. In *Dig. Tech. Papers Int'l Solid-State Circuits Conf. (ISSCC)*, pages 108–109, February 2004.

[152] M. Mori, M. Katsuno, S. Kasuga, T. Murata, and T. Yamaguchi. A 1/4in 2M pixel CMOS image sensor with 1.75 transistor/pixel. In *Dig. Tech. Papers Int'l Solid-State Circuits Conf. (ISSCC)*, pages 110–111, February 2004.

[153] K. Mabuchi, N. Nakamura, E. Funatsu, T. Abe, T. Umeda, T. Hoshino, R. Suzuki, and H. Sumi. CMOS image sensor using a floating diffusion driving buried photodiode. In *Dig. Tech. Papers Int'l Solid-State Circuits Conf. (ISSCC)*, pages 112–113, February 2004.

[154] M. Murakami, M. Masuyama, S. Tanaka, M. Uchida, K. Fujiwara, M Kojima, Y. Matsunaga, and S. Mayumi. 2.8 μm-Pixel Image Sensor vMaicovicon^{TM}. In *IEEE Workshop on Charge-Coupled Devices & Advanced Image Sensors*, pages 13–14, Karuizawa, Japan, June 2005.

[155] Y.C. Kim, Y.T. Kim, S.H. Choi, H.K. Kong, S.I. Hwang, J.H. Ko, B.S. Kim, T.Asaba, S.H. Lim, J.S. Hahn, J.H. Im, T.S. Oh, D.M. Yi, J.M. Lee, W.P. Yang, J.C. Ahn, E.S. Jung, and Y.H. Lee. 1/2-inch 7.2M Pixel CMOS Image Sensor with 2.25 μm Pixels Using 4-Shared Pixel Structure for Pixel-Level Summation. In *Dig. Tech. Papers Int'l Solid-State Circuits Conf. (ISSCC)*, pages 1994–2003, February 2006.

[156] S. Yoshihara, M. Kikuchi, Y. Ito, Y. Inada, S. Kuramochi, H. Wakabayashi, M. Okano, K. Koseki, H. Kuriyama, J. Inutsuka, A. Tajima, T. Nakajima, Y. Kudoh, F. Koga, Y. Kasagi, S. Watanabe, and T. Nomoto T. A 1/1.8-inch 6.4M Pixel 60 frames/s CMOS Image Sensor with Seamless Mode Change. In *Dig. Tech. Papers Int'l Solid-State Circuits Conf. (ISSCC)*, pages 1984–1993, February 2006.

[157] S. Yoshimura, T. Sugiyama, K. Yonemoto, and K. Ueda. A 48 kframe/s C-MOS image sensor for real-time 3-D sensing and motion detection. In *Dig. Tech. Papers Int'l Solid-State Circuits Conf. (ISSCC)*, pages 94–95, February 2001.

[158] J. Nakamura, B. Pain, T. Nomoto, T. Nakamura, and Eric R. Fossum. On-Focal-Plane Signal Processing for Current-Mode Active Pixel Sensors. *IEEE Trans. Electron Devices*, 44(10):1747–1758, October 1997.

[159] Y. Huanga and R.I. Hornsey. Current-mode CMOS image sensor using lateral bipolar phototransistors. *IEEE Trans. Electron Devices*, 50(12):2570–2573, December 2003.

[160] M.A. Szelezniak, G.W. Deptuch, F. Guilloux, S. Heini, and A. Himmi. Current Mode Monolithic Active Pixel Sensor with Correlated Double Sampling for Charged Particle Detection. *IEEE Sensors Journal*, 7(1):137–150, January 2007.

[161] L.G. McIlrath, V.S. Clark, P.K. Duane, R.D. McGrath, and W.D. Waskurak. Design and Analysis of a 512×768 Current-Mediated Active Pixel Array Image Sensor. *IEEE Trans. Electron Devices*, 44(10), 1997.

[162] F. Boussaid, A. Bermak, and A. Bouzerdoum. An ultra-low power operating technique for Mega-pixels current-mediated CMOS imagers. *IEEE Trans. Consumer Electron.*, 50(1):46–53, February 2004.

[163] D. Scheffer, B. Dierickx, and G. Meynants. Random Addressable 2048×2048 Active Pixel Image Sensor. *IEEE Trans. Electron Devices*, 44(10):1716–1720, October 1997.

[164] S.Kavadias, B. Dierickx, D. Scheffer, A. Alaerts, D. Uwaerts, and J. Bogaerts. A logarithmic response CMOS image sensor with on-chip calibration. *IEEE J. Solid-State Circuits*, 35(8):1146–1152, August 2000.

[165] M. Loose, K. Meier, and J. Schemmel. A self-calibrating single-chip CMOS camera with logarithmic response. *IEEE J. Solid-State Circuits*, 36(4):586–596, April 2001.

[166] Y. Oike, M. Ikeda, and K. Asada. High-Sensitivity and Wide-Dynamic-Range Position Sensor Using Logarithmic-Response and Correlation Circuit. *IEICE Trans. Electron.*, E85-C(8):1651–1658, August 2002.

[167] L.-W. Lai, C.-H. Lai, and Y.-C. King. A Novel Logarithmic Response CMOS Image Sensor with High Output Voltage Swing and In-Pixel Fixed-Pattern Noise Reduction. *IEEE Sensors Journal*, 4(1):122–126, February 2004.

[168] B. Choubey, S. Aoyoma, S. Otim, D. Joseph, and S. Collins. An Electronic-Calibration Scheme for Logarithmic CMOS Pixels. *IEEE Sensors Journal*, 6(4):950–956, August 2006.

[169] T. Kakumoto, S. Yano, M. Kusudaa, K. Kamon, and Y. Tanaka. Logarithmic conversion CMOS image sensor with FPN cancellation and integration circuits. *J. Inst. Image Information & Television Eng.*, 57(8):1013–1018, August 2003.

[170] J. Lazzaro, S. Ryckebuscha, M.A. Mahowald, and C. A. Mead. Winner-Take-All Networks of $O(n)$ Complexity. In D. Tourestzky, editor, *Advances in Neural Information Processing Systems*, volume 1, pages 703–711. Morgan Kaufmann, San Mateo, CA, 1988.

[171] S.-C. Liu, J. Karmer, G. Indiveri, T. Delbrük, and R. Douglas. *Analog VLSI Cicuits and Principles*. The MIT Press, Cambridge, MA, 2002.

[172] B.K.P. Horn. *Robot Vision*. The MIT Press, Cambridge, MA, 1986.

[173] E. Funatsu, Y. Nitta, Y. Miyake, T. Toyoda, J. Ohta, and K. Kyuma. An Artificial Retina Chip with Current-Mode Focal Plane Image Processing Functions. *IEEE Trans. Electron Devices*, 44(10):1777–1782, October 1997.

[174] E. Funatsu, Y. Nitta, J. Tanaka, and K. Kyuma. A 128 × 128 Pixel Artificial Retina LSI with Two-Dimensional Filtering Functions. *Jpn. J. Appl. Phys.*, 38(8B):L938–L940, August 1999.

[175] D. Marr. *Vision: A Computational Investigation into the Human Representation and Processing of Visual Information*. W. H. Freeman, 1983.

[176] X. Liu and A. El Gamal. Photocurrent Estimation for a Self-Reset CMOS Image Sensor. In *Proc. SPIE*, volume 4669, pages 304–312, San Jose, CA, 2002.

[177] K.P. Frohmader. A novel MOS compatible light intensity-to-frequency converter suited for monolithic integration. *IEEE J. Solid-State Circuits*, 17(3):588–591, June 1982.

[178] R. Müller. I²/L timing circuit for the 1 ms-10 s range. *IEEE J. Solid-State Circuits*, 12(2):139–143, April 1977.

[179] V. Brajovic and T. Kanade. A sorting image sensor: An example of massivley parallel intensity-to-time processing for low-latency computational sensors. In *Proc. IEEE Int'l Conf. Robotics & Automation*, pages 1638–1643, Minneapolis, MN, April 1996.

[180] M. Nagata, J. Funakoshi, and A. Iwata. A PWM Signal Processing Core Circuit Based on a Switched Current Integration Technique. *IEEE J. Solid-State Circuits*, 1998.

[181] M. Nagata, M. Homma, N. Takeda, T. Morie, and A. Iwata. A smart CMOS imager with pixel level PWM signal processing. In *Dig. Tech. Papers Symp. VLSI Circuits*, pages 141–144, June 1999.

[182] M. Shouho, K. Hashiguchi, K. Kagawa, and J. Ohta. A Low-Voltage Pulse-Width-Modulation Image Sensor. In *IEEE Workshop on Charge-Coupled Devices & Advanced Image Sensors*, pages 226–229, Karuizawa, Japan, June 2005.

[183] S. Shishido, I. Nagahata, T. Sasaki, K. Kagawa, M. Nunoshita, and J. Ohta. Demonstration of a low-voltage three-transistor-per-pixel CMOS imager

based on a pulse-width-modulation readout scheme employed with a one-transistor in-pixel comparator. In *Proc. SPIE*, San Jose, CA, 2007. Electronic Imaging.

[184] D. Yang, A. El Gamal, B. Fowler, and H. Tian. A 640 × 512 CMOS Image Sensor with Ultrawide Dynamic Range Floating-Point Pixel-Level ADC. *IEEE J. Solid-State Circuits*, 34(12):1821–1834, December 1999.

[185] S. Kleinfelder, S.-H. Lim, X. Liu, and A. El Gamal. A 10 000 Frames/s CMOS Digital Pixel Sensor. *IEEE J. Solid-State Circuits*, 36(12):2049–2059, December 2001.

[186] W. Biderman, A. El Gamal, S. Ewedemi, J. Reyneri, H. Tian, D. Wile, and D. Yang. A 0.18 μm High Dynamic Range NTSC/PAL Imaging System-on-Chip with Embedded DRAM Frame Buffer. In *Dig. Tech. Papers Int'l Solid-State Circuits Conf. (ISSCC)*, pages 212–213, 2003.

[187] J.G. Nicholls, A.R. Martin, B.G. Wallace, and P.A. Fuchs. *From Neuro to Brain*. Sinauer Associates, Inc., Sunderland, MA, 4th edition, 2001.

[188] W. Maass and C.M. Bishop, editors. *Pulsed Neural Networks*. The MIT Press, Cambridge, MA, 1999.

[189] T. Lehmann and R. Woodburn. Biologically-inspired learning in pulsed neural networks. In G. Cauwenberghs and M.A. Bayoumi, editors, *Learning on silicon: adaptive VLSI neural systems*, pages 105–130. Kluwer Academic Pub., Norwell, MA, 1999.

[190] K. Kagawa, K. Yasuoka, D. C. Ng, T. Furumiya, T. Tokuda, J. Ohta, and M. Nunoshita. Pulse-domain digital image processing for vision chips employing low-voltage operation in deep-submicron technologies. *IEEE Selcted Topic Quantum Electron.*, 10(4):816–828, July 2004.

[191] T. Hammadou. Pixel Level Stochastic Arithmetic for Intelligent Image Capture. In *Proc. SPIE*, volume 5301, pages 161–167, San Jose, CA, January 2004.

[192] W. Yang. A wide-dynamic-range, low-power photosensor array. In *Dig. Tech. Papers Int'l Solid-State Circuits Conf. (ISSCC)*, pages 230–231, February 1994.

[193] K. Kagawa, S. Yamamoto, T. Furumiya, T. Tokuda, M. Nunoshita, and J. Ohta. A pulse-frequency-modulation vision chip using a capacitive feedback reset with an in-pixel 1-bit image processing. In *Proc. SPIE*, volume 6068, pages 60680C–1–60680C–9, San Jose, January 2006.

[194] X. Wang, W. Wong, and R. Hornsey. A High Dynamic Range CMOS Image Sensor with Inpixel Light-to-Frequency Conversion. *IEEE Trans. Electron Devices*, 53(12):2988–2992, December 2006.

[195] T. Serrano-Gotarredona, A.G. Andreou, and B. Linares-Barranco. AER image filtering architecture for vision-processing systems. *IEEE Trans. Circuits & Systems I*, 46(9):1064–1071, September 1999.

[196] E. Culurciello, R. Etienne-Cummings, and K.A. Boahen. A Biomorphic Digital Image Sensor. *IEEE J. Solid-State Circuits*, 38(2):281–294, February 2003.

[197] T. Teixeira, A.G. Andreou, and E. Culurciello. An Address-Event Image Sensor Network. In *Int'l Symp. Circuits & Systems (ISCAS)*, pages 644–647, Kobe, Japan, May 2005.

[198] T. Teixeira, E. Culurciello, and A.G. Andreou. An Address-Event Image Sensor Network. In *Int'l Symp. Circuits & Systems (ISCAS)*, pages 4467–4470, Kos, Greece, May 2006.

[199] M.L. Simpson, G.S. Sayler, G. Patterson, D.E. Nivens, E.K. Bolton, J.M. Rochelle, and J.C. Arnott. An integrated CMOS microluminometer for low-level luminescence sensing in the bioluminescent bioreporter integrated circuit. *Sensors & Actuators B*, 72:134–140, 2001.

[200] E.K. Bolton, G.S. Sayler, D.E. Nivens, J.M. Rochelle, S. Ripp, and M.L. Simpson. Integratged CMOS photodetectors and signal processing for very low level chemical sensing with the bioluminescent bioreporter integrated circuits. *Sensors & Actuators B*, 85:179–185, 2002.

[201] J. Ohta, N. Yoshida, K. Kagawa, and M. Nunoshita. Proposal of Application of Pulsed Vision Chip for Retinal Prosthesis. *Jpn. J. Appl. Phys.*, 41(4B):2322–2325, April 2002.

[202] K. Kagawa, K. Isakari, T. Furumiya, A. Uehara, T. Tokuda, J. Ohta, and M. Nunoshita and. Pixel design of a pulsed CMOS image sensor for retinal prosthesis with digital photosensitivity control. *Electron. Lett.*, 39(5):419–421, May 2003.

[203] A. Uehara, K. Kagawa, T. Tokuda, J. Ohta, and M. Nunoshita. Back-illuminated pulse-frequency-modulated photosensor using a silicon-on-sapphire technology developed for use as an epi-retinal prosthesis device. *Electron. Lett.*, 39(15):1102–1104, July 2003.

[204] David C. Ng, K. Isakari, A. Uehara, K. Kagawa, T. Tokuda, J. Ohta, and M. Nunoshita. A study of bending effect on pulsed frequency modulation based photosensor for retinal prosthesis. *Jpn. J. Appl. Phys.*, 42(12):7621–7624, December 2003.

[205] T. Furumiya, K. Kagawa, A. Uehara, T. Tokuda, J. Ohta, and M. Nunoshita. 32 × 32-pixel pulse-frequency-modulation based image sensor for retinal prosthesis. *J. Inst. Image Information & Television Eng.*, 58(3):352–361, March 2004. In Japanese.

[206] K. Kagawa, N. Yoshida, T. Tokuda, J. Ohta, and M. Nunoshita. Building a Simple Model of A Pulse-Frequency-Modulation Photosensor and Demonstration of a 128 × 128-pixel Pulse-Frequency-Modulation Image Sensor Fabricated in a Standard 0.35-μm Complementary Metal-Oxide Semiconductor Technology. *Opt. Rev.*, 11(3):176–181, May 2004.

[207] A. Uehara, K. Kagawa, T. Tokuda, J. Ohta, and M. Nunoshita. A high-sensitive digital photosensor using MOS interface-trap charge pumping. *IEICE Electronics Express*, 1(18):556–561, December 2004.

[208] T. Furumiya, D. C. Ng, K. Yasuoka, K. Kagawa, T. Tokuda, M. Nunoshita, and J. Ohta. Functional verification of pulse frequency modulation-based image sensor for retinal prosthesis by *in vitro* electrophysiological experiments using frog retina. *Biosensors & Bioelectron.*, 21(7):1059–1068, January 2006.

[209] S. Yamamoto, K. Kagawa, T. Furumiya, T. Tokuda, M. Nunoshita, and J. Ohta. Prototyping and evaluation of a 32 × 32-pixel pulse-frequency-modulation vision chip with capacitive-feedback reset. *J. Inst. Image Information & Television Eng.*, 60(4):621–626, April 2006. In Japanese.

[210] T. Furumiya, S. Yamamoto, K. Kagawa, T. Tokuda, M. Nunoshita, and J. Ohta. Optimization of electrical stimulus pulse parameter for low-power operation of a retinal prosthetic device. *Jpn. J. Appl. Phys.*, 45(19):L505–L507, May 2006.

[211] D. C. Ng, T. Furumiya, K. Yasuoka, A. Uehara, K. Kagawa, T. Tokuda, M. Nunoshita, and J. Ohta. Pulse Frequency Modulation-based CMOS Image Sensor for Subretinal Stimulation. *IEEE Trans. Circuits & Systems II*, 53(6):487–491, June 2006.

[212] J. Ohta, T. Tokuda, K. Kagawa, T. Furumiya, A. Uehara, Y. Terasawa, M. Ozawa, T. Fujikado, and Y. Tano. Silicon LSI-Based Smart Stimulators for Retinal Prosthesis. *IEEE Eng. Medicine & Biology Magazine*, 25(5):47–59, October 2006.

[213] J. Deguchi, T. Watanabe, T. Nakamura, Y. Nakagawa, T. Fukushima, S. Jeoung-Chill, H. Kurino, T. Abe, M. Tamai, and M. Koyanagi. Three-Dimensionally Stacked Analog Retinal Prosthesis Chip. *Jpn. J. Appl. Phys.*, 43(4B):1685–1689, April 2004.

[214] D. Ziegler, P. Linderholm, M. Mazza, S. Ferazzutti, D. Bertrand, A.M. Ionescu, and Ph. Renaud. An active microphotodiode array of oscillating pixels for retinal stimulation. *Sensors & Actuators A*, 110:11–17, 2004.

[215] M. Mazza, P. Renaud, D.C. Bertrand, and A.M. Ionescu. CMOS Pixels for Subretinal Implantable Prothesis. *IEEE Sensors Journal*, 5(1):32–27, February 2005.

[216] M.L. Prydderch, M.J. French, K. Mathieson, C. Adams, D. Gunning, J. Laudanski, J.D. Morrison, A.R.Moodie, and J. Sinclair. A CMOS Active Pixel

Sensor for Retinal Stimulation. In *Proc. SPIE*, pages 606803–1–606803–9, San Jose, 2006. Electronic Imaging.

[217] S. Kagami, T. Komuro, and M. Ishikawa. A Software-Controlled Pixel-Level ADC Conversion Method for Digital Vision Chips. In *IEEE Workshop on Charge-Coupled Devices & Advanced Image Sensors*, Elmau, Germany, May 2003.

[218] J.B. Kuoa and S.-C. Lin. *Low-voltage SOI CMOS VLSI devices and circuits*. John Wiley & Sons, Inc., New York, NY, 2001.

[219] A. Afzalian and D. Flandre. Modeling of the bulk versus SOI CMOS performances for the optimal design of APS circuits in low-power low-voltage applications. *IEEE Trans. Electron Devices*, 2003.

[220] K. Senda, E. Fujii, Y. Hiroshima, and T. Takamura. Smear-less SOI image sensor. In *Tech. Dig. Int'l Electron Devices Meeting (IEDM)*, pages 369–372, 1986.

[221] K. Kioi, T. Shinozaki, S. Toyoyama, K. Shirakawa, K. Ohtake, and S. Tsuchimoto. Design and implementation of a 3D-LSI image sensing processor. *IEEE J. Solid-State Circuits*, 27(8):1130–1140, August 1992.

[222] V. Suntharalingam, R. Berger, J.A. Burns, C.K. Chen, C.L. Keast, J.M. Knecht, R.D. Lambert, K.L. Newcomb, D.M. O'Mara, D.D. Rathman, D.C. Shaver, A.M. Soares, C.N. Stevenson, B.M. Tyrrell, K. Warner, B.D. Wheeler, D.-R.W. Yost, and D.J. Young. Megapixel CMOS image sensor fabricated in three-dimensional integrated circuit technology. In *Dig. Tech. Papers Int'l Solid-State Circuits Conf. (ISSCC)*, pages 356–357, February 2005.

[223] X. Zheng, S. Seshadri, M. Wood, C. Wrigley, and B. Pain. Process and Pixels for High Performance Imager in SOI-CMOS Technology. In *IEEE Workshop on Charge-Coupled Devices & Advanced Image Sensors*, Elmau, Germany, May 2003.

[224] B. Pain. Fabrication and Initial Results for a Back-illuminated Monolithic APS in a Mixed SOI/Bulk CMOS Technology. In *IEEE Workshop on Charge-Coupled Devices & Advanced Image Sensors*, pages 102–104, Karuizawa, Japan, June 2005.

[225] Y.S. Cho, H. Takano, K. Sawada, M. Ishida, and S.Y. Choi. SOI CMOS Image Sensor with Pinned Photodiode on Handle Wafer. In *IEEE Workshop on Charge-Coupled Devices & Advanced Image Sensors*, pages 105–108, Karuizawa, Japan, June 2005.

[226] S. Iwabuchi, Y. Maruyama, Y. Ohgishi, M. Muramatsu, N. Karasawa, and T. Hirayama. A Back-Illuminated High-Sensitivity Small-Pixel Color CMOS Image Sensor with Flexible Layout of Metal Wiring. In *Dig. Tech. Papers Int'l Solid-State Circuits Conf. (ISSCC)*, pages 1171–1178, February 2006.

[227] H. Yamamoto, K. Taniguchi, and C. Hamaguchi. High-sensitivity SOI MOS photodetector with self-amplification. *Jpn. J. Appl. Phys.*, 35(2B):1382–1386, February 1996.

[228] W. Zhang, M. Chan, S.K.H. Fung, and P.K. Ko. Performance of a CMOS compatible lateral bipolar photodetector on SOI substrate. *IEEE Electron Device Lett.*, 19(11):435–437, November 1998.

[229] W. Zhang, M. Chan, and P.K. Ko. Performance of the floating gate/body tied NMOSFET photodetector on SOI substrate. *IEEE Trans. Electron Devices*, 47(7):1375–1384, July 2000.

[230] C. Xu, W. Zhang, and M. Chan. A low voltage hybrid bulk/SOI CMOS active pixel image sensor. *IEEE Electron Device Lett.*, 22(5):248–250, May 2001.

[231] C. Xu, C. Shen, W. Wu, and M. Chan. Backside-Illuminated Lateral PIN Photodiode for CMOS Image Sensor on SOS Substrate. *IEEE Trans. Electron Devices*, 52(6):1110–1115, June 2005.

[232] T. Ishikawa, M. Ueno, Y. Nakaki, K. Endo, Y. Ohta, J. Nakanishi, Y. Kosasayama, H. Yagi, T. Sone, and M. Kimata. Performance of 320 × 240 Uncooled IRFPA with SOI Diode Detectors. In *Proc. SPIE*, volume 4130, pages 152–159, 2000.

[233] A.G. Andreou, Z.K. Kalayjian, A. Apsel, P.O. Pouliquen, R.A. Athale, G. Simonis, and R. Reedy. Silicon on sapphire CMOS for optoelectronic microsystems. *IEEE Circuits & Systems Magazine*, 2001.

[234] E. Culurciello and A.G. Andreou. 16 × 16 pixel silicon on sapphire CMOS digital pixel photosensor array. *Electron. Lett.*, 40(1):66–68, January 2004.

[235] A. Fish, E. Avner, and O. Yadid-Pecht. Low-power global/rolling shutter image sensors in silicon on sapphire technology. In *Int'l Symp. Circuits & Systems (ISCAS)*, pages 580–583, Kobe, Japan, May 2005.

[236] S.D. Gunapala, S.V. Bandara, J.K. Liu, Sir B. Rafol, and J.M. Mumolo. 640 × 512 Pixel Long-Wavelength Infrared Narrowband, Multiband, and Broadband QWIP Focal Plane Arrays. *IEEE Trans. Electron Devices*, 50(12):2353–2360, December 2003.

[237] M. Kimata. Infrared Focal Plane Arrays. In H. Baltes, W. Gopel, and J. Hesse, editors, *Sensors Update*, volume 4, pages 53–79. Wiley-VCH, 1998.

[238] C.-C. Hsieh, C.-Y. Wu, F.-W. Jih, and T.-P. Sun. Focal-Plane-Arrays and C-MOS Readout Techniques of Infrared Imaging Systems. *IEEE Trans. Circuits & Systems Video Tech.*, 7(4):594–605, August 1997.

[239] M. Kimata. Silicon infrared focal plane arrays. In M. Henini and M. Razeghi, editors, *Handbook of Infrared Detection Technologies*, pages 352–392. Elsevier Science Ltd., 2002.

[240] E. Kasper and K. Lyutovich, editors. *Properties of Silicon Germanium and SiGe:Carbon*. INSPEC, The Institute of Electrical Engineers, London, UK, 2000.

[241] T. Tokuda, Y. Sakano, K. Kagawa, J. Ohta, and M. Nunoshita. Backside-hybirid photodetector for trans-chip detection of NIR light. In *IEEE Workshop on Charge-Coupled Devices & Advanced Image Sensors*, Elmau, Germany, May 2003.

[242] T. Tokuda, D. Mori, K. Kagawa, M. Nunoshita, and J. Ohta. A CMOS image sensor with eye-safe detection function using backside carrier injection. *J. Inst. Image Information & Television Eng.*, 60(3):366–372, March 2006. In Japanese.

[243] J.A. Burns, B.F. Aull, C.K. Chen, C.-L. Chen, C.L. Keast, J.M. Knecht, V. Suntharalingam, K. Warner, P.W. Wyatt, and D.-R.W. Yost. A Wafer-Scale 3-D Circuit Integration Technology. *IEEE Trans. Electron Devices*, 53(10):2507–2516, October 2006.

[244] M. Koyanagi, T. Nakamura, Y. Yamada, H. Kikuchi, T. Fukushima, T. Tanaka, and H. Kurino. Three-Dimensional Integration Technology Based on Wafer Bonding with Vertical Buried Interconnections. *IEEE Trans. Electron Devices*, 53(11):2799–2808, November 2006.

[245] A. Iwata, M. Sasaki, T. Kikkawa, S. Kameda, H. Ando, K. Kimoto, D. Arizono, and H. Sunami. A 3D integration scheme utilizing wireless interconnections for implementing hyper brains. In *Dig. Tech. Papers Int'l Solid-State Circuits Conf. (ISSCC)*, pages 262–597, February 2005.

[246] N. Miura, D. Mizoguchi, M. Inoue, K. Niitsu, Y. Nakagawa, M. Tago, M. Fukaishi, T. Sakurai, and T. Kuroda. A 1Tb/s 3W inductive-coupling transceiver for inter-chip clock and data link. In *Dig. Tech. Papers Int'l Solid-State Circuits Conf. (ISSCC)*, pages 1676–1685, February 2006.

[247] E. Culurciello and A.G. Andreou. Capacitive Coupling of Data and Power for 3D Silicon-on-Insulator VLSI. In *Int'l Symp. Circuits & Systems (ISCAS)*, pages 4142–4145, Kobe, Japan, May 2005.

[248] D.A.B. Miller, A. Bhatnagar, S. Palermo, A. Emami-Neyestanak, and M.A. Horowitz. Opportunities for Optics in Integrated Circuits Applications. In *Dig. Tech. Papers Int'l Solid-State Circuits Conf. (ISSCC)*, pages 86–87, February 2005.

[249] K. W. Lee, T. Nakamura, K. Sakuma, K.T. Park, H. Shimazutsu, N. Miyakawa, K.Y. Kim, H. Kurino, and M. Koyanagi. Development of Three-Dimensional Integration Technology for Highly Parallel Image-Processing Chip. *Jpn. J. Appl. Phys.*, 39(4B):2473–2477, April 2000.

[250] M. Koyanagi, Y. Nakagawa, K.W. Lee, T. Nakamura, Y. Yamada, K. Inamura, K.-T. Park, and H. Kurino. Neuromorphic Vision Chip Fabricated Using

Three-Dimensional Integration Technology. In *Dig. Tech. Papers Int'l Solid-State Circuits Conf. (ISSCC)*, February 2001.

[251] S. Lombardo, S. U. Campisano, G. N. van den Hoven, and A. Polman. Erbium in oxygen-doped silicon: Electroluminescence. *J. Appl. Phys.*, 77(12):6504–6510, December 1995.

[252] M. A. Green, J. Zhao, A. Wang, P. J. Reece, and M. Gal. Efficient silicon light-emitting diodes. *Nature*, 412:805–808, 2001.

[253] L.W. Snyman, M. du Plessis, E. Seevinck, and H. Aharoni. An Efficient Low Voltage, High Frequency Silicon CMOS Light Emitting Device and Electro-Optical Interface. *IEEE Trans. Electron Devices*, 20(12):614–617, December 1999.

[254] M. du Plessis, H. Aharoni, and L.W. Snyman. Spatial and intensity modulation of light emission from a silicon LED matrix. *IEEE Photon. Tech. Lett.*, 14(6):768–770, June 2002.

[255] J. Ohta, K. Isakari, H. Nakayama, K. Kagawa, T. Tokuda, and M. Nunoshita. An image sensor integrated with light emitter using BiCMOS process. *J. Inst. Image Information & Television Eng.*, 57(3):378–383, March 2003. In Japanese.

[256] H. Aharoni and M. du Plessis. Low-operating-voltage integrated silicon light-emitting devices. *IEEE J. Quantum Electron.*, 40(5):557–563, May 2004.

[257] M. Sergio and E. Charbon. An intra-chip electro-optical channel based on CMOS single photon detectors. In *Tech. Dig. Int'l Electron Devices Meeting (IEDM)*, December 2005.

[258] J. R. Haynes and W. C. Westphal. Radiation Resulting from Recombination of Holes and Electrons in Silicon. *Phys. Rev.*, 101(6):1676–1678, March 1956.

[259] S. Maëstre and P. Magnan. Electroluminescence and Impact Ionization in CMOS Active Pixel Sensors. In *IEEE Workshop on Charge-Coupled Devices & Advanced Image Sensors*, Elmau, Germany, May 2003.

[260] K.M. Findlater, D.Renshaw, J.E.D. Hurwitz, R.K. Henderson, T.E.R. Biley, S.G. Smith, M.D. Purcell, and J.M. Raynor. A CMOS Image Sensor Employing a Double Junction Photodiode. In *IEEE Workshop on Charge-Coupled Devices & Advanced Image Sensors*, pages 60–63, Lake Tahoe, NV, June 2001.

[261] D.L. Gilblom, S.K. Yoo, and P. Ventura. Operation and performance of a color image sensor with layered photodiodes. In *Proc. SPIE*, volume 5074 of *SPIE AeroSense*, pages 5210–14–27, Orlando, FL, April 2003.

[262] R.B. Merrill. Color Separation in an Active Pixel Cell Imaging Array Using a Triple-Well Structure. US Patent 5,965,875, 1999.

[263] T. Lulé, B. Schneider, and M. Böhm. Design and Fabrication of a High-Dynamic-Range Image Sensor in TFA Technology. *IEEE J. Solid-State Circuits*, 34(5):704–711, May 1999.

[264] M. Sommer, P. Rieve, M. Verhoeven, M. Böhm, B. Schneider, B. van Uffel, and F. Librecht. First Multispectral Diode Color Imager with Three Color Recognition and Color Memory in Each Pixel. In *IEEE Workshop on Charge-Coupled Devices & Advanced Image Sensors*, pages 187–190, Karuizawa, Japan, June 1999.

[265] H. Steibig, R.A. Street, D. Knipp, M. Krause, and J. Ho. Vertically integrated thin-film color sensor arrays for advanced sensing applications. *Appl. Phys. Lett.*, 88:013509, 2006.

[266] Y. Maruyama, K. Sawada, H. Takao, and M. Ishida. The fabrication of filterless fluorescence detection sensor array using CMOS image sensor technique. *Sensors & Actuators A*, 128:66–70, 2006.

[267] Y. Maruyama, K. Sawada, H. Takao, and M. Ishida. A novel filterless fluorescence detection sensor for DNA analysis. *IEEE Trans. Electron Devices*, 53(3):553–558, March 2006.

[268] P. Catrysse, B. Wandell, and A. El Gamal. An integrated color pixel in 0.18μm CMOS technology. In *Tech. Dig. Int'l Electron Devices Meeting (IEDM)*, pages 24.4.1–24.4.4, December 2001.

[269] K. Sasagawa, K. Kusawake, K. Kagawa, J. Ohta, and M. Nunoshita. Optical transmission enhancement for an image sensor with a sub-wavelength aperture. In *Int'l Conf. Optics-Photonics Design & Fabrication (ODF)*, pages 163–164, November 2002.

[270] L.W. Barnes, A. Dereux, and T.W. Ebbesen. Surface plasmon subwavelength optics. *Nature*, 2003.

[271] Y. Inaba, M. Kasano, K. Tanaka, and T. Yamaguchi. Degradation-free MOS image sensor with photonic crystal color filter. *IEEE Electron Device Lett.*, 27(6):457–459, June 2006.

[272] H.A. Bethe. Theory of Diffraction by Small Holes. *Phys. Rev.*, 66(7-8):163–182, October 1944.

[273] T. Thio, K.M. Pellerin, R.A. Linke, H.J. Lezec, and T.W. Ebbessen. Enhanced light transmission through a single subwavelength aperture. *Opt. Lett.*, 26, December 2001.

[274] H. Raether. *Surface Plasmons on Smooth and Rough Surfaces and on Gratings*. Springer-Verlag, Berlin, Germany, 1988.

[275] B.J. Hosticka, W. Brockherde, A. Bussmann, T. Heimann, R. Jeremias, A. Kemna, C. Nitta, and O. Schrey. CMOS imaging for automotive applications. *IEEE Trans. Electron Devices*, 50(1):173–183, January 2003.

[276] J. Hynecek. Impactron — A New Solid State Image Intensifier. In *IEEE Workshop on Charge-Coupled Devices & Advanced Image Sensors*, pages 197–200, Lake Tahoe, NV, June 2001.

[277] S. Ohta, H. Shibuya, I. Kobayashi, T. Tachibana, T. Nishiwaki, and J. Hynecek. Characterization Results of 1k × 1k Charge Multiplying CCD Image Sensor. In *Proc. SPIE*, volume 5301, pages 99–108, San Jose, CA, January 2004.

[278] H. Eltoukhy, K. Salama, and A.E. Gamal. A 0.18-μm CMOS bioluminescence detection lab-on-chip. *IEEE J. Solid-State Circuits*, 2006.

[279] H. Jia and P.A. Abshire. A CMOS image sensor for low light applications. In *Int'l Symp. Circuits & Systems (ISCAS)*, pages 1651–1654, Kos, Greece, May 2006.

[280] B. Fowler, M.D. Godfrey, J. Balicki, and J. Canfield. Low Noise Readout using Active Reset for CMOS APS. In *Proc. SPIE*, volume 3965, pages 126–135, San Jose, CA, January 2000.

[281] I. Takayanagi, Y. Fukunaga, T. Yoshida, and J. Nakamura. A Four-Transistor Capacitive Feedback Reset Active Pixel and its Reset Noise Reduction Capability. In *IEEE Workshop on Charge-Coupled Devices & Advanced Image Sensors*, pages 118–121, Lake Tahoe, NV, June 2001.

[282] B. Pain, T.J. Cunningham, B. Hancock, G. Yang, S. Seshadri, and M. Ortiz. Reset noise suppression in two-dimensional CMOS photodiode pixels through column-based feedback-reset. In *Tech. Dig. Int'l Electron Devices Meeting (IEDM)*, pages 809–812, December 2002.

[283] S. Kleinfelder. High-speed, high-sensitivity, low-noise CMOS scientific image sensors. In *Proc. SPIE*, volume 5247 of *Microelectronics: Design, Technology, and Packaging*, pages 194–205, December 2003.

[284] T. J. Cunningham, B. Hancock, C. Sun, G. Yang, M. Oritz, C. Wrigley, S. Seshadri, and B. Pain. A Two-Dimensional Array Imager Demonstrting Active Reset Suppression of kTC-Noise. In *IEEE Workshop on Charge-Coupled Devices & Advanced Image Sensors*, Elmau, Germany, May 2003.

[285] Y. Chen and S. Kleinfelder. CMOS active pixel sensor achieving 90 dB dyanmic range with colum-level active reset. In *Proc. SPIE*, volume 5301, pages 438–449, San Jose, CA, 2004.

[286] K.-H. Lee and E. Yoon. A CMOS Image Sensor with Reset Level Control Using Dynamc Reset Current Source for Noise Suppression. In *Dig. Tech. Papers Int'l Solid-State Circuits Conf. (ISSCC)*, page 114, February 2004.

[287] L. Kozlowski, G. Rossi, L. Blanquart, R. Marchesini, Y. Huang, G. Chow, and J. Richardson. A Progressive 1920 × 1080 Imaging System-on-Chip for HDTV Cameras. In *Dig. Tech. Papers Int'l Solid-State Circuits Conf. (ISSCC)*, pages 358–359, February 2005.

[288] J. Yang, K.G. Fife, L. Brooks, C.G. Sodini, A. Betts, P. Mudunuru, and H.-S. Lee. A 3MPixel Low-Noise Flexible Architecture CMOS Image Sensor. In *Dig. Tech. Papers Int'l Solid-State Circuits Conf. (ISSCC)*, pages 2004–2013, February 2006.

[289] T.G. Etoh, D. Poggemann, G. Kreider, H. Mutoh, A.J.P. Theuwissen, A. Ruckelshausen, Y. Kondo, H. Maruno, K. Takubo, H. Soya, K. Takehara, T. Okinaka, and Y. Takano. An image sensor which captures 100 consecutive frames at 1000000 frames/s. *IEEE Trans. Electron Devices*, 50(1):144–151, January 2003.

[290] A.I. Krymski, N.E. Bock, N. Tu, D. Van Blerkom, and E.R. Fossum. A High-Speed, 240-Frames/s, 4.1-Mpixel CMOS Sensor. *IEEE Trans. Electron Devices*, 50(1):130–135, January 2003.

[291] A.I. Krymski and N. Tu. A 9-V/Lux-s 5000-frames/s 512 ×512 CMOS sensor. *IEEE Trans. Electron Devices*, 50(1):136–143, January 2003.

[292] Y. Nitta, Y. Muramatsu, K. Amano, T. Toyama, J. Yamamoto, K. Mishina, A. Suzuki, T. Taura, A. Kato, M. Kikuchi, Y. Yasui, H. Nomura, and N. Fukushima. High-Speed Digital Double Sampling with Analog CDS on Column Parallel ADC Architecture for Low-Noise Active Pixel Sensor. In *Dig. Tech. Papers Int'l Solid-State Circuits Conf. (ISSCC)*, pages 2024–2031, February 2006.

[293] A.I. Krymski and K. Tajima. CMOS Image Sensor with integrated 4Gb/s Camera Link Transmitter. In *Dig. Tech. Papers Int'l Solid-State Circuits Conf. (ISSCC)*, pages 2040–2049, February 2006.

[294] M. Furuta, T. Inoue, Y. Nishikawa, and S. Kawahito. A 3500fps High-Speed CMOS Image Sensor with 12b Column-Parallel Cyclic A/D Converters. In *Dig. Tech. Papers Symp. VLSI Circuits*, pages 21–22, June 2006.

[295] T. Inoue, S. Takeuchi, and S. Kawhito. CMOS active pixel image sensor with in-pixel CDS for high-speed cameras. In *Proc. SPIE*, volume 5301, pages 2510–257, San Jose, CA, January 2004.

[296] S. Lauxtermann, G. Israel, P. Seitz, H. Bloss, J. Ernst, H. Firla, and S. Gick. A mega-pixel high speed CMOS sensor with sustainable Gigapixel/s readout rate. In *IEEE Workshop on Charge-Coupled Devices & Advanced Image Sensors*, pages 48–51, Lake Tahoe, NV, June 2001.

[297] N. Stevanovic, M. Hillebrand, B.J. Hosticka, and A. Teuner. A CMOS image sensor for high-speed imaging. In *Dig. Tech. Papers Int'l Solid-State Circuits Conf. (ISSCC)*, pages 104–105, February 2000.

[298] N. Bock, A. Krymski, A. Sarwari, M. Sutanu, N. Tu, K. Hunt, M. Cleary, N. Khaliullin, and M. Brading. A wide-VGA CMOS image sensor with global shutter and extended dyamic range. In *IEEE Workshop on Charge-Coupled*

Devices & Advanced Image Sensors, pages 222–225, Karuizawa, Japan, June 2005.

[299] G. Yang and T. Dosluoglu. Ultra High Light Shutter Rejection Ratio Snap-shot Pixel Image Sensor ASIC for Pattern Recoginition. In *IEEE Workshop on Charge-Coupled Devices & Advanced Image Sensors*, pages 161–164, Karuizawa, Japan, June 2005.

[300] D.H. Hubel. *Eye, Brain, and Vision*. Scientific American Library, New York, NY, 1987.

[301] M. Mase, S. Kawahito, M. Sasaki, Y. Wakamori, and M. Furuta. A Wide Dynamic Range CMOS Image Sensor with Multiple Exposure-Time Signal Outputs and 12-bit Column-Parallel Cyclic AD Converters. *IEEE J. Solid-State Circuits*, 40(12):2787–2795, December 2005.

[302] Y. Wang, S.L. Barna, S. Campbell, and E.R. Fossum. A high dynamic range CMOS APS image sensor. In *IEEE Workshop on Charge-Coupled Devices & Advanced Image Sensors*, pages 137–140, Lake Tahoe, NV, June 2001.

[303] D. Yang and A. El Gamal. Comparative analysis of SNR for image sensors with widened dynamic range. In *Proc. SPIE*, volume 3649, pages 197–211, San Jose, CA, February 1999.

[304] R. Hauschild, M. Hillebrand, B.J. Hosticka, J. Huppertz, T. Kneip, and M. Schwarz. A CMOS image sensor with local brightness adaptation and high intrascene dynamic range. In *Proc. European Solid-State Circuits Conf. (ESSCIRC)*, pages 308–311, September 1998.

[305] O. Schrey, R. Hauschild, B.J. Hosticka, U. Iurgel, and M. Schwarz. A locally adaptive CMOS image sensor with 90 dB dynamic range. In *Dig. Tech. Papers Int'l Solid-State Circuits Conf. (ISSCC)*, pages 310–311, February 1999.

[306] S.L. Barna, L.P. Ang, B. Mansoorian, and E.R. Fossum. A low-light to sun-light, 60 frames/s, 80 kpixel CMOS APS camera-on-a-chip with 8b digital output. In *IEEE Workshop on Charge-Coupled Devices & Advanced Image Sensors*, pages 148–150, Karuizawa, Japan, June 1999.

[307] S. Lee and K. Yang. High dynamic-range CMOS image sensor cell based on self-adaptive photosensing operation. *IEEE Trans. Electron Devices*, 53(7):1733–1735, July 2006.

[308] S. Chamberlain and J.P.Y. Lee. A novel wide dynamic range silicon photode-tector and liner imaging array. *IEEE J. Solid-State Circuits*, 19(1), February 1984.

[309] G.G. Storm, J.E.D. Hurwitz, D. Renshawa, K.M. FIndlater, R.K. Hender-son, and M.D. Purcell. High dynamic range imaging using combined linear-logarithmic response from a CMOS image sensor. In *IEEE Workshop on Charge-Coupled Devices & Advanced Image Sensors*, Elmau, Germany, May 2003.

[310] K. Hara, H. Kubo, M. Kimura, F. Murao, and S. Komori. A Linear-Logarithmic CMOS Sensor with Offset Calibration Using an Injected Charge Signal. In *Dig. Tech. Papers Int'l Solid-State Circuits Conf. (ISSCC)*, 2005.

[311] S. Decker, D. McGrath, K. Brehmer, and C.G.Sodini. A 256 × 256 CMOS imaging array with wide dynamic range pixels and column-parallel digital output. *IEEE J. Solid-State Circuits*, 33(12):2081–2091, December 1998.

[312] Y. Muramatsu, S. Kurosawa, M. Furumiya, H. Ohkubo, and Y. Nakashiba. A Signal-Processing CMOS Image Sensor using Simple Analog Operation. In *Dig. Tech. Papers Int'l Solid-State Circuits Conf. (ISSCC)*, pages 98–99, February 2001.

[313] V. Berezin, I. Ovsiannikov, D. Jerdev, and R. Tsai. Dynamic Range Enlargement in CMOS Imagers with Buried Photodiode. In *IEEE Workshop on Charge-Coupled Devices & Advanced Image Sensors*, Elmau, Germany, 2003.

[314] S. Sugawa, N. Akahane, S. Adachi, K. Mori, T. Ishiuchi, and K. Mizobuchi. A 100 dB dyamic range CMOS image sensor using a lateral overflow integration capacitor. In *Dig. Tech. Papers Int'l Solid-State Circuits Conf. (ISSCC)*, pages 352–353, February 2005.

[315] N. Akahane, R. Ryuzaki, S. Adachi, K. Mizobuchi, and S. Sugawa. A 200dB Dynamic Range Iris-less CMOS Image Sensor with Lateral Overflow Integration Capacitor using Hybrid Voltage and Current Readout Operation. In *Dig. Tech. Papers Int'l Solid-State Circuits Conf. (ISSCC)*, February 2006.

[316] N. Akahane, S. Sugawa, S. Adachi, K. Mori, T. Ishiuchi, and K. Mizobuchi. A sensitivity and linearity improvement of a 100-dB dynamic range CMOS image sensor using a lateral overflow integration capacitor. *IEEE J. Solid-State Circuits*, 41(4):851–858, April 2006.

[317] M. Ikeba and K. Saito. CMOS-Image Sensor with PD-Capacitance Modulation using Negative Feedback Resetting. *J. Inst. Image Information & Television Eng.*, 60(3):384, March 2006.

[318] K. Kagawa, Y. Adachi, Y. Nose, H. Takashima, K. Tani, A. Wada, M. Nunoshita, and J. Ohta. A wide-dynamic-range CMOS imager by hybrid use of active and passive pixel sensors. In *IS&T SPIE Annual Symposium Electronic Imaging*, [6501-18], San Jose, CA, January 2007.

[319] S.T. Smith, P. Zalud, J. Kalinowski, N.J. McCaffrey, P.A. Levine, and M.L. Lin. BLINC: a 640 × 480 CMOS active pixel vide camera with adaptive digital processing, extended optical dynamic range, and miniature form factor. In *Proc. SPIE*, volume 4306, pages 41–49, January 2001.

[320] H. Witter, T. Walschap, G. Vanstraelen, G. Chapinal, G. Meynants, and B. Dierickx. 1024 × 1280 pixel dual shutter APS for industrial vision. In *Proc. SPIE*, volume 5017, pages 19–23, 2003.

[321] O. Yadid-Pecht and E.R. Fossum. Wide intrascene dynamic range CMOS APS using dual sampling. *IEEE Trans. Electron Devices*, 44(10):1721–1723, October 1997.

[322] O. Schrey, J. Huppertz, G. Filimonovic, A. Bussmann, W. Brockherde, and B.J. Hosticka. A 1 K × 1 K high dynamic range CMOS image sensor with on-chip programmable region-of-interest. *IEEE J. Solid-State Circuits*, 37(7):911–915, September 2002.

[323] K. Mabuchi, N. Nakamura, E. Funatsu, T. Abe, T. Umeda, T. Hoshino, R. Suzuki, and H. Sumi. CMOS image sensor using a floating diffusion driving buried photodiode. In *Dig. Tech. Papers Int'l Solid-State Circuits Conf. (ISSCC)*, pages 112–516, February 2004.

[324] M. Sasaki, M. Mase, S. Kawahito, and Y. Tadokoro. A Wide Dyamic Range CMOS Image Sensor wit Integration of Short-Exposure-Time Signals. In *IEEE Workshop on Charge-Coupled Devices & Advanced Image Sensors*, Elmau, Germany, 2003.

[325] M. Sasaki, M. Mase, S. Kawahito, and Y. Tadokoro. A wide dynamic range CMOS image sensor with multiple short-time exposures. In *Proc. IEEE Sensors*, volume 2, pages 967–972, October 2004.

[326] M. Schanz, C. Nitta, A. Bußmann, B.J. Hosticka, and R.K. Wertheimer. A high-dynamic-range CMOS image sensor for automotive applications. *IEEE J. Solid-State Circuits*, 35(7):932–938, September 2000.

[327] B. Schneider, H. FIsher, S. Benthien, H. Keller, T. Lulé, P. Rieve, M. SOmmer, and J. Schulte M. Böhm. TFA Image Sensors: From the One Transistor Cell to a Locally Adaptive High Dynamic Range Sensor. In *Tech. Dig. Int'l Electron Devices Meeting (IEDM)*, pages 209–212, December 1997.

[328] T. Hamamoto and K. Aizawa. A computational image sensor with adaptive pixel-based integration time. *IEEE J. Solid-State Circuits*, 36(4):580–585, April 2001.

[329] T. Yasuda, T. Hamamoto, and K. Aizawa. Adaptive-integration-time image sensor with real-time reconstruction function. *IEEE Trans. Electron Devices*, 50(1):111–120, January 2003.

[330] O. Yadid-Pecht and A. Belenky. In-pixel autoexposure CMOS APS. *IEEE J. Solid-State Circuits*, 38(8):1425–1428, August 2003.

[331] P.M. Acosta-Serafini, I. Masaki, and C.G. Sodini. A 1/3" VGA linear wide dynamic range CMOS image sensor implementing a predictive multiple sampling algorithm with overlapping integration intervals. *IEEE J. Solid-State Circuits*, 39(9):1487–1496, September 2004.

[332] T. Anaxagoras and N.M. Allinson. High dynamic range active pixel sensor. In *Proc. SPIE*, volume 5301, pages 149–160, San Jose, CA, January 2004.

[333] D. Stoppa, A. Simoni, L. Gonzo, M. Gottardi, and G.-F. Dalla Betta. Novel CMOS image sensor with a 132-dB dynamic range. *IEEE J. Solid-State Circuits*, 37(12):1846–1852, December 2002.

[334] J. Döge, G. Shöneberg, G.T. Streil, and A. König. An HDR CMOS Image Sensor with Spiking Pixels, Pixel-Level ADC, and Linear Characteristics. *IEEE Trans. Circuits & Systems II*, 49(2):155–158, February 2002.

[335] S. Chen, A. Bermak, and F. Boussaid. A compact reconfigurable counter memory for spiking pixels. *IEEE Electron Device Lett.*, 27(4):255–257, April 2006.

[336] K. Oda, H. Kobayashi, K. Takemura, Y. Takeuchi, and T. Yamda. The development of wide dynamic range iamge sensor. *ITE Tech. Report*, 27(25):17–20, 2003. In Japanese.

[337] S. Ando and A. Kimachi. Time-Domain Correlation Image Sensor: First CMOS Realization of Demodulation Pixels Array. In *IEEE Workshop on Charge-Coupled Devices & Advanced Image Sensors*, pages 33–36, Karuizawa, Japan, June 1999.

[338] T. Spirig, P. Seitz, O. Vietze, and F. Heitger. The lock-in CCD-two-dimensional synchronous detection of light. *IEEE J. Quantum Electron.*, 31(9):1705–1708, September 1995.

[339] T. Spirig, M. Marley, and P. Seitz. The multitap lock-in CCD with offset subtraction. *IEEE Trans. Electron Devices*, 44(10):1643–1647, October 1997.

[340] R. Lange, P. Seitz, A. Biber, and S. Lauxtermann. Demodulation Pixels in CCD and CMOS Technologies for Time-Of-Flight Ranging. In *Proc. SPIE*, pages 177–188, San Jose, CA, January 2000.

[341] R. Lange and P. Seitz. Solid-state time-of-flight range camera. *IEEE J. Quantum Electron.*, 37(3):390–397, March 2001.

[342] A. Kimachi, T. Kurihara, M. Takamoto, and S. Ando. A Novel Range Finding Systems Using Correlation Image Sensor. *Trans. IEE Jpn.*, 121-E(7):367–375, July 2001.

[343] K. Yamamoto, Y. Oya, K. Kagawa, J. Ohta, M. Nunoshita, and K. Watanabe. Demonstration of a freqency-demodulation CMOS image sensor and its improvement of image quality. In *IEEE Workshop on Charge-Coupled Devices & Advanced Image Sensors*, Elmau, Germany, June 2003.

[344] S. Ando and A. Kimachi. Correlation Image Sensor: Two-Dimensional Matched Detection of Amplitude-Modulated Light. *IEEE Trans. Electron Devices*, 50(10):2059–2066, October 2003.

[345] R. Miyagawa and T. Kanade. CCD-based range-finding sensor. *IEEE Trans. Electron Devices*, 44(10):1648–1652, October 1997.

[346] B. Buxbaum, R. Chwarte, and T. Ringbeck. PMD-PLL: Receiver structure for incoherent communication and ranging systems. In *Proc. SPIE*, volume 3850, pages 116–127, 1999.

[347] S.-Y. Ma and L.-G. Chen. A single-chip CMOS APS camera with direct frame difference output. *IEEE J. Solid-State Circuits*, 34(10):1415–1418, October 1999.

[348] J. Ohta, K. Yamamoto, T. Hirai, K. Kagawa, M. Nunoshita, M. Yamada, Y. Yamasaki, S. Sughishita, and K. Watanabe. An image sensor with an in-pixel demodulation function for detecting the intensity of a modulated light signal. *IEEE Trans. Electron Devices*, 50(1):166–172, January 2003.

[349] K. Yamamoto, Y. Oya, K. Kagawa, J. Ohta, M. Nunoshita, and K. Watanabe. Improvement of demodulated image quality in a demodulated image sensor. *J. Inst. Image Information & Television Eng.*, 57(9):1108–1114, September 2003. In Japanese.

[350] Y. Oike, M. Ikeda, and K. Asada. A 120 × 110 position sensor with the capability of sensitive and selective light detection in wide dynamic range for robust active range finding. *IEEE J. Solid-State Circuits*, 39(1):246–251, January 2004.

[351] Y. Oike, M. Ikeda, and K. Asada. Pixel-Level Color Demodulation Image Sensor for Support of Image Recognition. *IEICE Trans. Electron.*, E87-C(12):2164–2171, December 2004.

[352] K. Yamamoto, Y. Oya, K. Kagawa, M. Nunoshita, J. Ohta, and K. Watanabe. A 128 × 128 Pixel CMOS Image Sensor with an Improved Pixel Architecture for Detecting Modulated Light Signals. *Opt. Rev.*, 13(2):64–68, April 2006.

[353] B. Buttgen, F. Lustenberger, and P. Seitz. Demodulation Pixel Based on Static Drift Fields. *IEEE Trans. Electron Devices*, 53(11):2741–2747, November 2006.

[354] A. Kimachi, H. Ikuta, Y. Fujiwara, and H. Matsuyama. Spectral matching imager using correlation image sensor and AM-coded multispectral illuminatin. In *Proc. SPIE*, volume 5017, pages 128–135, Santa Clara, CA, January 2003.

[355] K. Asashi, M. Takahashi, K. Yamamoto, K. Kagawa, and J. Ohta. Application of a demodulated image sensos for a camera sysystem suppressing saturation. *J. Inst. Image Information & Television Eng.*, 60(4):627–630, April 2006. In Japanese.

[356] Y. Oike, M. Ikeda, and K. Asada. Design and implementation of real-time 3-D image sensor with 640 × 480 pixel resolution. *IEEE J. Solid-State Circuits*, 39(4):622–628, April 2004.

[357] B. Aull, J. Burns, C. Chen, B. Felton, H. Hanson, C. Keast, J. Knecht, A. Loomis, M. Renzi, A. Soares, V. Suntharalingam, K. Warner, D. Wolfson, D. Yost, and D. Young. Laser Radar Imager Based on 3D Integration of

Geiger-Mode Avalanche Photodiodes with Two SOI Timing Circuit Layers. In *Dig. Tech. Papers Int'l Solid-State Circuits Conf. (ISSCC)*, pages 1179–1188, February 2006.

[358] L. Viarani, D. Stoppa, L. Gonzo, M. Gottardi, and A. Simoni. A CMOS Smart Pixel for Active 3-D Vision Applications. *IEEE Sensors Journal*, 4(1):145–152, February 2004.

[359] D. Stoppa, L. Viarani, A. Simoni, L. Gonzo, M. Malfatti, and G. Pedretti. A 50 × 50-pixel CMOS sensor for TOF-based real time 3D imaging. In *IEEE Workshop on Charge-Coupled Devices & Advanced Image Sensors*, pages 230–233, Karuizawa, Japan, June 2005.

[360] R. Jeremias, W. Brockherde, G. Doemens, B. Hosticka, L. Listl, and P. Mengel. A CMOS photosensor array for 3D imaging using pulsed laser. In *Dig. Tech. Papers Int'l Solid-State Circuits Conf. (ISSCC)*, pages 252–253, February 2001.

[361] O. Elkhalili, O.M. Schrey, P. Mengel, M. Petermann, W. Brockherde, and B.J. Hosticka. A 4 × 64 pixel CMOS image sensor for 3-D measurement applications. *IEEE J. Solid-State Circuits*, 30(7):1208–1212, February 2004.

[362] T. Sawasa, T Ushinaga, I.A. Halin, S. Kawahito, M. Homma, and Y. Maeda. A QVGA-size CMOS time-of-flight range image sensor with background ligh charge draining structure. *ITE Tech. Report*, 30(25):13–16, March 2006. In Japanese.

[363] T. Moller, H. Kraft, J. Frey, M. Albrecht, and R. Lange. Robust 3D Measurement with PMD Sensors. In *Proc. 1st Range Imaging Research Day at ETH Zurich*, page Supplement to the Proceedings, Zurich, 2005.

[364] T. Kahlmann, F. Remondino, and H. Ingensand. Calibration for increased accuracy of the range imaging camera SwissRanger^{TM}. In *Int'l Arch. Photogrammetry, Remote Sensing & Spatial Information Sci.*, volume XXXVI part 5, pages 136–141. Int'l Soc. Photogrammetry & Remote Sensing (IS-PRS) Commission V Symposium, September 2006.

[365] M. Lehmann, T. Oggier, B. Büttgen, Chr. Gimkiewicz, M. Schweizer, R. Kaufmann, F. Lustenberger, and N. Blanc. Smart pixels for future 3D-TOF sensors. In *IEEE Workshop on Charge-Coupled Devices & Advanced Image Sensors*, pages 193–196, Karuizawa, Japan, June 2005.

[366] S.B. Gokturk, H. Yalcin, and C. Bamji. A Time-of-Flight Depth Sensor — System Description, Issues and Solutions. In *Conf. Computer Vision & Pattern Recognition Workshop (CVPR)*, pages 35–44, Washington, DC, June 2004.

[367] B. Pain, L. Matthies, B. Hancock, and C.Sun. A compact snap-shot range-imaging receiver. In *IEEE Workshop on Charge-Coupled Devices & Advanced Image Sensors*, pages 234–237, Karuizawa, Japan, June 2005.

[368] T. Kato, S. Kawahito, K. Kobayashi, H. Sasaki, T. Eki, and T. Hisanaga. A Bioncular CMOS Range Image Sensor with Bit-Serial Block-Parallel Interface Using Cyclic Pipelined ADCś. In *Dig. Tech. Papers Symp. VLSI Circuits*, pages 270–271, Honolulu, Hawaii, June 2002.

[369] S. Kakehi, S. Nagao, and T. Hamamoto. Smart Image Sensor with Binocular PD Array for Tracking of a Moving Object and Depth Estimation. In *Int'l Symposium Intelligent Signal Processing & Communication Systems (ISPAC-S)*, pages 635–638, Awaji, Japan, December 2003.

[370] R.M. Philipp and R. Etienne-Cummings. Single chip stero imager. In *Int'l Symp. Circuits & Systems (ISCAS)*, pages 808–811, 2003.

[371] R.M. Philipp and R. Etienne-Cummings. A 128 × 128 33mW 30frames/s single-chip stereo imager. In *Dig. Tech. Papers Int'l Solid-State Circuits Conf. (ISSCC)*, pages 2050–2059, February 2006.

[372] A. Gruss, L.R. Carley, and T. Kanade. Integrated Sensor and Range-Finding Analog Signal Processor. *IEEE J. Solid-State Circuits*, 26(3):184–191, March 1991.

[373] Y. Oike, M. Ikeda, and K. Asada. A CMOS Image Sensor for High-Speed Active Range Finding Using Column-Parallel Time-Domain ADC and Position Encoder. *IEEE Trans. Electron Devices*, 50(1):152–158, January 2003.

[374] Y. Oike, M. Ikeda, and K. Asada. A 375 × 365 high-speed 3-D range-finding image sensor using row-parallel search architecture and multisampling technique. *IEEE J. Solid-State Circuits*, 40(2):444–453, February 2005.

[375] T. Sugiyama, S. Yoshimura, R. Suzuki, and H. Sumi. A 1/4-inch QVGA color imaging and 3-D sensing CMOS sensor with analog frame memory. In *Dig. Tech. Papers Int'l Solid-State Circuits Conf. (ISSCC)*, pages 434–479, February 2002.

[376] H. Miura, H. Ishiwata, Y. Lida, Y. Matunaga, S. Numazaki, A. Morisita, N. Umeki, and M. Doi. 100 frame/s CMOS active pixel sensor for 3D-gesture recognition system. In *Dig. Tech. Papers Int'l Solid-State Circuits Conf. (ISSCC)*, pages 142–143, February 1999.

[377] M.D. Adams. Coaxial Range Measurement — Current Trends for Mobile Robotic Applications. *IEEE Sensors Journal*, 2(1):2–13, February 2002.

[378] D. Vallancourt and S.J. Daubert. Applications of current-copier circuits. In C. Toumazou, F.J. Lidgey, and D.G. Haigh, editors, *Analogue IC design: the current-mode approach*, IEE Circuts and Systems Series 2, chapter 14, pages 515–533. Peter Peregrinus Ltd., London, UK, 1990.

[379] K. Aizawa, K. Sakaue, and Y. Suenaga, editors. *Image Processig Technologies, Alogorithms, Sensors, and Applications*. Mracel Dekker, Inc., New York, NY, 2004.

[380] A. Yokota, T. Yoshida, H. Kashiyama, and T. Hamamoto. High-speed Sensing System for Depth Estimation Based on Depth-from-Focus by Using Smart Imager. In *Int'l Symp. Circuits & Systems (ISCAS)*, pages 564–567, Kobe, Japan, May 2005.

[381] V. Milirud, L. Fleshel, W. Zhang, G. Jullien, and O. Yadid-Pecht. A WIDE DYNAMIC RANGE CMOS ACTIVE PIXEL SENSOR WITH FRAME D-IFFERENCE. In *Int'l Symp. Circuits & Systems (ISCAS)*, pages 588–591, Kobe, Japan, May 2005.

[382] U. Mallik, M. Clapp, E. Choi, G. Cauwenbergs, and R. Etienne-Cummings. Temporal Change Threshold Detectin Imager. In *Dig. Tech. Papers Int'l Solid-State Circuits Conf. (ISSCC)*, 2005.

[383] V. Brajovic and T. Kanade. Computational Sensor for Visual Tracking with Attention. *IEEE J. Solid-State Circuits*, 33(8):1199–1207, August 1998.

[384] J. Akita, A. Watanabe, O. Tooyama, M. Miyama, and M. Yoshimoto. An Image Sensor with Fast ObjectsṔositions Extraction Function. *IEEE Trans. Electron Devices*, 50(1):184–190, January 2003.

[385] R. Etienne-Cummings, J. Van der Spiegel, P. Mueller, and M.-Z. Zhang. A foveated silicon retina for two-dimensional tracking. *IEEE Trans. Circuits & Systems II*, 47(6):504–517, June 2000.

[386] R.D. Burns, J. Shah, C. Hong, S. Pepić, J.S. Lee, R.I. Hornsey, and P. Thomas. Object Location and Centroiding Techniques with CMOS Active Pixel Sensors. *IEEE Trans. Electron Devices*, 50(12):2369–2377, December 2003.

[387] Y. Sugiyama, M. Takumi, H. Toyoda, N. Mukozaka, A. Ihori, T. Kurashina, Y. Nakamura, T. Tonbe, and S. Mizuno. A High-Speed CMOS Image Sensor with Profile Data Acquiring Function. *IEEE J. Solid-State Circuits*, 40(12):2816–2823, December 2005.

[388] T.G. Constandinou and C. Toumazou. A Micropower Centroiding Vision Processor. *IEEE J. Solid-State Circuits*, 41(6):1430–1443, June 2006.

[389] H. Oku, Teodorus, K. Hashimotoa, and M. Ishikawa. High-speed Focusing of Cells Using Depth-From-Diffraction Method. In *Proc. IEEE Int'l Conf. Robotics & Automation (ICRA)*, pages 3636–3641, Orlando, FL, May 2006.

[390] M.F. Land and D.-E. Nilsson. *Animal Eyes*. Oxford University Press, Oxford, UK, 2002.

[391] E. Hecht. *Optics*. Addison-Wesley Pub. Co., Reading, MA, 2nd edition, 1987.

[392] B.A. Wandell. *Foundations of Vision*. Sinauer Associates, Inc., Sunderland, MA, 1995.

[393] R. Wodnicki, G.W. Roberts, and M.D. Levine. A Log-Polar Image Sensor Fabricated in a Standard 1.2-μm ASIC CMOS Process. *IEEE J. Solid-State Circuits*, 32(8):1274–1277, August 1997.

[394] F. Pardo, B. Dierickx, and D. Scheffer. CMOS foveated image sensor: signal scaling and small geometry effects. *IEEE Trans. Electron Devices*, 44(10):1731–1737, October 1997.

[395] F. Pardo, B. Dierickx, and D. Scheffer. Space-variant nonorthogonal structure CMOS image sensor design. *IEEE J. Solid-State Circuits*, 33(6):842–849, June 1998.

[396] F. Saffih and R. Hornsey. Pyramidal Architecture for CMOS Image Sensor. In *IEEE Workshop on Charge-Coupled Devices & Advanced Image Sensors*, Elmau, Germany, May 2003.

[397] K. Yamazawa, Y. Yagi, and M. Yachida. Ominidirectional Imaging with Hyperboloidal Projection. In *Proc. IEEE/RSJ Int'l Conf. Intelligent Robots & Systems*, pages 1029–1034, Yokohama, Japan, July 1993.

[398] J. Ohta, H. Wakasa, K. Kagawa, M. Nunoshita, M. Suga, M. Doi, M. Oshiro, K. Minato, and K. Chihara. A CMOS image sensor for Hyper Omni Vision. *Trans. Inst. Electrical Eng. Jpn., E*, 123-E(11):470–476, November 2003.

[399] J. Tanidaa, T. Kumagai, K. Yamada, S. Miyatakea, K. Ishida, T. Morimoto, N. Kondou, D. Miyazaki, and Yoshiki Ichioka. Thin observation module by bound optics (TOMBO): concept and experimental verification. *Appl. Opt.*, 40(11):1806–1813, April 2001.

[400] S. Ogata, J. Ishida, and H. Koshi. Optical sensor array in an artificial compound eye. *Opt. Eng.*, 33:3649–3655, November 1994.

[401] J.S. Sanders and C.E. Halford. Design and analysis of apposition compound eye optical sensros. *Opt. Eng.*, 34(1):222–235, January 1995.

[402] K. Hamanaka and H. Koshi. An artificial compound eye using a microlens array and its application to scale-invariant processing. *Opt. Rev.*, 3(4):265–268, 1996.

[403] J. Duparré, P. Dannberg, P. Schreiber, A. Bräuer, and A. Tünnermann. Microoptically fabricated artificial apposition compound eye. In *Proc. SPIE*, volume 5301, pages 25–33, San Jose, CA, January 2004.

[404] R. Hornsey, P. Thomas, W. Wong, S. Pepic, K. Yip, and R. Kishnasamy. Electronic compound-eye image sensor: construction and calibration. In *Proc. SPIE*, volume 5301, pages 13–24, San Jose, CA, January 2004.

[405] J. Tanida, R. Shogenji, Y. Kitamura, K. Yamada, M. Miyamoto, and S. Miyatake. Imaging with an integrated compound imaging system. *Opt. Express*, 11(18):2109–2117, September 2003.

[406] R. Shogenji, Y. Kitamura, K. Yamada, S. Miyatake, and J. Tanida. Multispectral imaging using compact compound optics. *Opt. Express*, 12(8):1643–1655, April 2004.

[407] S. Miyatake, R. Shogenji, M. Miyamoto, K. Nitta, and J. Tanida. Thin observation module by bound optics (TOMOBO) with color filters. In *Proc. SPIE*, volume 5301, pages 7–12, San Jose, CA, January 2004.

[408] A.G. Andreou and Z.K. Kalayjian. Polarization Imaging: Principles and Integrated Polarimeters. *IEEE Sensors Journal*, 2(6):566–576, 2002.

[409] M. Nakagawa. Ubiquitous Visible Light Communications. *IEICE Trans. on Communications*, J88-B(2):351–359, February 2005.

[410] PhaseSpace, Inc. Phase Space motion digitizer. www.phasespace.com/.

[411] PhoeniX Technologies. The Visualeyes System. ptiphoenix.com/.

[412] D. J. Moore, R. Want, B. L. Harrison, A. Gujar, and K. Fishkin. Implementing Phicons: Combining Computer Vision with InfraRed Technology for Interactive Physical Icons. In *Proc. ACM Symposium on User Interface Software and Technology (UIST)*, pages 67–68, 1999.

[413] J. Rekimoto and K. Nagao. The World through the Computer: Augmented Interaction with Real World Environment. In *Proc. ACM Symposium on User Interface Software and Technology (UIST)*, pages 29–36, 1995.

[414] N. Matsushita, D. Hihara, T. Ushiro, S. Yoshimura, and J. Rekimoto. ID Cam; A smart camera for scene capturing and ID recognition. *J. Informatin Process. Soc. Jpn.*, 43(12):3664–3674, December 2002.

[415] N. Matsushita, D. Hihara, T. Ushiro, S. Yoshimura, J. Rekimoto, and Y. Yamamoto. ID CAM: A Smart Camera for Scene Capturing and ID Recognition. *Proc. IEEE & ACM Int'l Sympo. Mixed & Augmented Reality*, page 227, 2003.

[416] H. Itoh, K. Kosugi, X. Lin, Y. Nakamura, T. Nishimura, K. Takiizawa, and H. Nakashima. Spatial Optical point-to-point communication system for indoor locarion-based information services. In *Proc. ICO Int'l Conf. Optics & Photonics in Technology Frontier*, July 2004.

[417] X. Lin and H. Itoh. Wireless Personal Information Terminal for Indoor Spatial Optical Communication System Using a Modified DataSlim2. *Opt. Rev.*, 10(3):155–160, May–Jun 2003.

[418] K. Kagawa, Y. Maeda, K. Yamamoto, Y. Masaki, J. Ohta, and M. Nunoshita. Optical navigation: a ubiquitous visual remote-control station for home information appliances. In *Proc. Optics Japan*, pages 112–113, 2004. In Japanese.

[419] K.Kagawa, K. Yamamoto, Y. Maeda, Y. Miyake, H. Tanabe, Y. Masaki, M. Nunoshita, and J. Ohta. "Opto-Navi," a Multi-Purpose Visual Remote Controller of Home Information Appliances Using a Custom CMOS Image Sensor. *Forum on Info. Tech. Lett.*, 4:229–232, April 2005. In Japanese.

[420] K. Kagawa, R. Danno, K. Yamamoto, Y. Maeda, Y. Miyake, H. Tanabe, Y. Masaki, M. Nunoshita, and J. Ohta. Demonstration of mobile visual remote

controller "OptNavi" system using home network. *J. Inst. Image Information & Television Eng.*, 60(6):897–908, June 2006.

[421] www.dlna.org/.

[422] www.echonet.gr.jp/.

[423] www.upnp.org/.

[424] www.havi.org/.

[425] www.irda.org/.

[426] www.bluetooth.com/.

[427] Y. Oike, M. Ikeda, and K. Asada. A smart image sensor with high-speed feeble ID-beacon detection for augmented reality system. In *Proc. European Solid-State Circuits Conf.*, pages 125–128, September 2003.

[428] Y. Oike, M. Ikeda, and K. Asada. Smart Image Sensor with High-speed High-sensitivity ID Beacon Detection for Augmented Reality System. *J. Inst. Image Information & Television Eng.*, 58(6):835–841, June 2004. In Japanese.

[429] K. Yamamoto, K. Kagawa, Y. Maeda, Y. Miyake, H. Tanabe, Y. Masaki, M. Nunoshita, and J. Ohta. An Opt-Navi system using a custom CMOS image sensor with a function of reading multiple region-of-interests. *J. Inst. Image Information & Television Eng.*, 59(12):1830–1840, December 2005. In Japanese.

[430] K. Yamamoto, Y. Maeda, Y. Masaki, K. Kagawa, M. Nunoshita, and J. Ohta. A CMOS image sensor with high-speed readout of mulitple region-of-interests for an Opto-Navigation system. In *Proc. SPIE*, volume 5667, pages 90–97, San Jose, CA, January 2005.

[431] K. Yamamoto, Y. Maeda, Y. Masaki, K. Kagawa, M. Nunoshita, and J. Ohta. A CMOS image sensor for ID detection with high-speed readout of multiple region-of-interests. In *IEEE Workshop on Charge-Coupled Devices & Advanced Image Sensors*, pages 165–168, Karuizawa, Japan, June 2005.

[432] J.R. Barry. *Wireless infrared communications*. Kluwer Academic Publishers, New York, NY, 1994.

[433] www.victor.co.jp/pro/lan/index.html.

[434] J.M. Kahn, R. You, P. Djahani, A. G.Weisbin, B. K. Teik, and A. Tang. Imaging diversity receivers for high-speed infrared wireless communication. *IEEE Commun. Mag.*, 36(12):88–94, December 1998.

[435] D.C. O'Brien, G.E. Faulkner, E.B. Zyambo, K. Jim, D.J. Edwards, P. Stavrinou, G. Parry, J. Bellon, M.J. Sibley, V.A. Lalithambika, V.M. Joyner, R.J. Samsudin, D.M. Holburn, and R.J. Mears. Integrated Transceivers for Optical Wireless Communications. *IEEE Selcted Topic Quantum Electron.*, 11(1):173–183, Jan–Feb. 2005.

[436] K. Kagawa, T. Nishimura, T. Hirai, J. Ohta, M. Nunoshita, M. Yamada, Y. Yamasaki, S. Sughishita, and K. Watanabe. A vision chip with a focused high-speed read-out mode for optical wireless LAN. In *Int'l Topical Meeting on Optics in Computing*, pages 183–185, Engelberg, Switzerland, April 2002.

[437] T. Nishimura, K. Kagawa, T. Hirai, J. Ohta, M. Nunoshita, M. Yamada, Y. Yamasaki, S. Sughishita, and K. Watanabe. Design and Fabrication fo High-speed BiCMOS Image Sensor with Focused Read-out Mode for Optical Wireless LAN. In *Tech. Digest 7th Optoelectronics & Communications Conf. (OECC2002)*, pages 212–213, July 2002.

[438] K. Kagawa, T. Nishimura, J. Ohta, M. Nunoshita, Y. Yamasaki, M. Yamada, S. Sughishita, and K. Watanabe. An optical wirelesds LAN system based on a MEMS beam steerer and a vision chip. In *Int'l Conf. Optics-photonics Design & Fabrication (ODF)*, pages 135–136, November 2002.

[439] K. Kagawa, T. Nishimura, H. Asazu, T. Kawakami, J. Ohta, M. Nunoshita, Y. Yamasaki, and K. Watanabe. A CMOS Image Sensor Working As High-Speed Photo Receiver as Well as a Position Sensor for Indoor Optical Wireless LAN Systems. In *Proc. SPIE*, volume 5017, pages 86–93, Santa Clara, CA, January 2003.

[440] K. Kagawa, T. Kawakami, H. Asazu, T. Nishimura, J. Ohta, M. Nunoshita, and K. Watanabe. An image sensor based optical receiver fabricated in a standard 0.35 μm CMOS technology for mobile applications. In *IEEE Workshop on Charge-Coupled Devices & Advanced Image Sensors*, Elmau, Germany, May 2003.

[441] K. Kagawa, T. Nishimura, T.Hirai, Y. Yamasaki, J. Ohta, M. Nunoshita, and K. Watanabe. Proposal and Preliminary Experiments of Indoor optical wireless LAN based on a CMOS image sensor with a high-speed readout function enabling a low-power compact module with large downlink capacity. *IEICE Trans. Commun.*, E86-B(5):1498–1507, May 2003.

[442] K. Kagawa, T. Kawakami, H. Asazu, T. Ikeuchi, A. Fujiuchi, J. Ohta, M. Nunoshita, and K. Watanabe. An indoor optical wireless LAN system with a CMOS-image-sensor-based photoreceiver fabricated in a 0.35-μm CMOS technology. In *Int'l Topical Meeting Optics in Computing*, pages 90–91, Engelberg, Switzerland, April 2004.

[443] K. Kagawa, H. Asazu, T. Kawakami, T. Ikeuchi, A. Fujiuchi, J. Ohta, M. Nunoshita, and K. Watanabe. Design and fabrication of a photoreceiver for a spatially optical communication using an image sensor. *J. Inst. Image Information & Television Eng.*, 58(3):334–343, March 2004. In Japanese.

[444] A. Fujiuchi, T. Ikeuchi, K. Kagawa, J. Ohta, and M. Nunoshita. Free-space wavelength-division-multiplexing optical communications using a multichannel photoreceiver. In *Int'l Conf. Optics & Photonics in Technology Frontier (ICO)*, pages 480–481, Chiba, Japan, July 2004.

[445] K. Kagawa, H. Asazu, T. Ikeuchi, Y. Maeda, J. Ohta, M. Nunoshita, and K. Watanabe. A 4-ch 400-Mbps image-sensor-based photo receiver for indoor optical wireless LANs. In *Tech. Dig. 9th Optoelectronics & Communications Conf. (OECC2004)*, pages 822–823, Yokohama, Japan, June 2004.

[446] K. Kagawa, T. Ikeuchi, J. Ohta, and M. Nunoshita. An Image sensor with a photoreceiver function for indoor optical wireless LANs fabricated in 0.8-μm BiCMOS technology. In *Proc. IEEE Sensors*, page 288, Vienna, Austria, October 2004.

[447] B.S. Leibowitz, B.E. Boser, and K.S.J. Pister. A 256-Element CMOS Imaging Receiver for Free-Space Optical Communication. *IEEE J. Solid-State Circuits*, 40(9):1948–1956, September 2005.

[448] M. Last, B.S. Leibowitz, B. Cagdaser, A. Jog, L. Zhou, B.E. Boser, and K.S.J. Pister. Toward a wireless optical communication link between two small unmanned aerial vehicles. In *Int'l Symp. Circuits & Systems (ISCAS)*, volume 3, pages 930–933, May 2003.

[449] B. Eversmann, M. Jenkner, F. Hofmann, C. Paulus, R. Brederlow, B. Holzapfl, P. Fromherz, M. Merz, M. Brenner, M. Schreiter, R. Gabl, K. Plehnert, M. Steinhauser, G. Eckstein, D. Schmitt-Landsiedel, and R. Thewes. A 128 × 128 CMOS biosensor array for extracellular recording of neural activity. *IEEE J. Solid-State Circuits*, 38(12):2306–2317, December 2003.

[450] U. Lu, B. Hu, Y. Shih, C.Wu, and Y. Yang. The design of a novel complementary metal oxide semiconductor detection system for biochemical luminescence. *Biosensors Bioelectron.*, 19(10):1185–1191, 2004.

[451] H. Ji, P.A. Abshire, M. Urdaneta, and E. Smela. CMOS contact imager for monitoring cultured cells. In *Int'l Symp. Circuits & Systems (ISCAS)*, pages 3491–3495, Kobe, Japan, May 2005.

[452] K. Sawada, T. Ohshina, T. Hizawa, H. Takao, and M. Ishida. A novel fused sensor for photo- and ion-sensing. *Sensors & Actuators B*, 106:614–618, 2005.

[453] H. Ji, D. Sander, A. Haas, and P.A. Abshire. A CMOS contact imager for locating individual cells. In *Int'l Symp. Circuits & Systems (ISCAS)*, pages 3357–3360, Kos, Greece, May 2006.

[454] J. C. Jackson, D. Phelan, A. P. Morrison, M. Redfern, and A. Mathewson. Characterization of Geiger Mode Avalanche Photodiodes for Fluorescence Decay Measurements. In *SPIE, Photodetector Materials and Devices VII*, volume 4650, San Jose, CA, January 2002.

[455] D. Phelan, J. C. Jackson, R. M. Redfern, A. P. Morrison, and A. Mathewson. Geiger Mode Avalanche Photodiodes for Microarray Systems. In *SPIE, Biomedical Nanotechnology Architectures and Applications*, volume 4626, San Jose, CA, January 2002.

[456] S. Bellis, J.C. Jackson, and A. Mathewson. Towards a Disposable in vivo Miniature Implantable Fluorescence Detector. In *SPIE, Optical Fibers and Sensors for Medical Diagnostics and Treatment Applications VI*, volume 6083, 2006.

[457] T. Tokuda, A. Yamamoto, K. Kagawa, M. Nunoshita, and J. Ohta. A CMOS image sensor with optical and potential dual imaging function for on-chip bioscientific applications. *Sensors & Actuators A*, 125(2):273–280, February 2006.

[458] T. Tokuda, I. Kadowaki, .K Kagawa, M. Nunoshita, and J. Ohta. A new imaging scheme for on-chip DNA spots with optical / potential dual-image CMOS sensor in dry situation. *Jpn. J. Appl. Phys.*, 46(4B):2806–2810, April 2007.

[459] T. Tokuda, K. Tanaka, M. Matsuo, K. Kagawa, M. Nunoshita, and J. Ohta. Optical and electrochemical dual-image CMOS sensor for on-chip biomolecular sensing applications. *Sensors & Actuators A*, 135(2):315–322, April 2007.

[460] D. C. Ng, T. Tokuda, A. Yamamoto, M. Matsuo, M. Nunoshita, H. Tamura, Y. Ishikawa, S. Shiosaka, and J. Ohta. A CMOS Image Sensor for On-chip *in vitro* and *in vivo* Imaging of the Mouse Hippocampus. *Jpn. J. Appl. Phys.*, 45(4B):3799–3806, April 2006.

[461] D. C. Ng, H. Tamura, T. Tokuda, A. Yamamoto, M. Matsuo, M. Nunoshita, Y. Ishikawa, S. Shiosaka, and J. Ohta. Real Time *In vivo* Imaging and Measurement of Serine Protease Activity in the Mouse Hippocampus Using a Dedicated CMOS Imaging Device. *J. Neuroscience Methods*, 156(1-2):23–30, September 2006.

[462] D. C. Ng, T. Tokuda, A. Yamamoto, M. Matsuo, M. Nunoshita, H. Tamura, Y. Ishikawa, S. Shiosaka, and J. Ohta. On-chip biofluorescence imaging inside a brain tissue phantom using a CMOS image sensor for *in vivo* brain imaging verification. *Sensors & Actuators B*, 119(1):262–274, November 2006.

[463] D.C. Ng, T. Nakagawa, T. Tokuda, M. Nunoshita, H. Tamura, Y. Ishikawa, S. Shiosaka, and J. Ohta. Development of a Fully Integrated Complementary Metal-Oxide Semiconductor Image Sensor-based Device for Real-time *In vivo* Fluorescence Imaging inside the Mouse Hippocampus. *Jpn. J. Appl. Phys.*, 46(4B):2811–2819, April 2007.

[464] T. Sakata, M. Kamahori, and Y. Miyahara. Immobilization of oligonucleotide probes on Si_3N_4 surface and its application to genetic field effect transistor. *Mater. Sci. Eng.*, C24:827–832, 2004.

[465] M. Urdaneta, M. Christophersen, E. Smela, S.B. Prakash, N. Nelson, and P. Abshire. Cell Clinics Technology Platform for Cell-Based Sensing. In *IEEE/NLM Life Science Systems & Applications Workshop*, Bethesda, MD, July 2006.

[466] K. Hashimoto, K. Ito, and Y. Ishimori. Microfabricated disposable DNA sensor for detection of hepatitis B virus DNA. *Sensors & Actuators B*, 46:220–225, 1998.

[467] K. Dill, D.D. Montgomery, A.L. Ghindilis, and K.R. Schwarzkopf. Immunoassays and sequence-specific DNA detection on a microchip using enzyme amplified electrochemical detection. *J. Biochem. Biophys. Methods*, 2004.

[468] H. Miyahara, K. Yamashita, M. Takagi, H. Kondo, and S. Takenaka. Electrochemical array (ECA) as and integrated multi-electrode DNA sensor. *T. IEE Jpn.*, 121-E:187–191, 2004.

[469] A. Frey, M. Schienle, C. Paulus, Z. Jun, F. Hofmann, P. Schindler-Bauer, B. Holzapfl, M. Atzesberger, G. Beer, M. Frits, T. Haneder, H.-C. Hanke, and R. Thewes. A digital CMOS DNA chip. In *Int'l Symp. Circuits & Systems (ISCAS)*, pages 2915–2918, Kobe, Japan, May 2005.

[470] R.D. Frostig, editor. *In Vivo Optical Imaging of Brain Function*. CRC Press, Boca Raton, FL, 2002.

[471] A. W. Toga and J. C. Mazziota. *Brain Mapping: the Methods*. Academic Press, New York, NY, 2nd edition, 2002.

[472] C. Shimizu, S. Yoshida, M. Shibata, K. Kato, Y. Momota, K. Matsumoto, T. Shiosaka, R. Midorikawa, T. Kamachi, A. Kawabe, and S. Shiosaka. Characterization of recombinant and brain neuropsin, a plasticity-related serine protease. *J Biol Chem.*, 273:11189–11196, 1998.

[473] G. Iddan, G. Meron, A. Glukhovsky, and P. Swain. Wireless capsule endoscopy. *Nature*, 405(6785):417, 2000.

[474] R. Eliakim. Esophageal capsule endoscopy (ECE): four case reports which demonstrate the advantage of bi-directional viewing of the esophagus. In *Int'l Conf. Capsule Endoscopy*, pages 109–110, Florida, 2004.

[475] S. Itoh and S. Kawahito. Frame Sequential Color Imaging Using CMOS Image Sensors. In *ITE Annual Convention*, pages 21–2, 2003. In Japanese.

[476] S. Itoh, S. Kawahito, and S. Terakawa. A 2.6mW 2fps QVGA CMOS One-chip Wireless Camera with Digital Image Transmission Function for Capsule Endoscopes. In *Int'l Symp. Circuits & Systems (ISCAS)*, pages 3353–3356, Kos, Greece, May 2006.

[477] X. Xie, G. Li, X. Chen, X. Li, and Z. Wang. A Low-Power Digital IC Design Inside the Wireless Endoscopic Capsule. *IEEE J. Solid-State Circuits*, 40(11):2390–2400, 2006.

[478] S. Kawahito, M. Yoshida, M. Sasaki, K, Umehara, D. Miyazaki, Y. Tadokoro, K. Murata, S. Doushou, and A. Matsuzawa. A CMOS Image Sensor

with Analog Two-Dimensional DCT-Based Compression Circuits for one-chip Cameras. *IEEE J. Solid-State Circuits*, 32(12):2030–2041, December 1997.

[479] K. Aizawa, Y. Egi, T. Hamamoto, M. Hatoria, M. Abe, H. Maruyama, and H. Otake. Computational image sensor for on sensor compression. *IEEE Trans. Electron Devices*, 44(10):1724–1730, October 1997.

[480] Z. Lin, M.W. Hoffman, W.D. Leon-Salas, N. Schemm, and S. Balkr. A C-MOS Image Sensor for Focal Plane Decomposition. In *Int'l Symp. Circuits & Systems (ISCAS)*, pages 5322–5325, Kobe, Japan, May 2005.

[481] A. Bandyopadhyay, J. Lee, R. Robucci, and P. Hasler. A 8 μW/frame 104 $\times$ 128 CMOS imager front end for JPEG Compression. In *Int'l Symp. Circuits & Systems (ISCAS)*, pages 5318–5312, Kobe, Japan, May 2005.

[482] T.B. Tang, E. A. Johannesen, L. Wang, A. Astaras, M. Ahmadian, A. F. Murray, J.M Cooper, S.P. Beaumont, B.W. Flynn, and D.R.S. Cumming. Toward a Miniature Wireless Integrated Multisensor Microsystem for Industrial and Biomedical Applications. *IEEE Sensors Journal*, 2:628–635, 2002.

[483] A. Astaras, M. Ahmadian, N. Aydin, L. Cui, E. Johannessen, T.-B. Tang., L. Wang, T. Arslan, S.P. Beaumont, B.W. Flynn, A.F. Murray, S.W. Reid, P. Yam, J.M. Cooper, and D.R.S. Cumming. A miniature integrated electronics sensor capsule for real-time monitoring of the gastrointestinal tract (IDEAS). In *Int'l Conf. Biomedical Eng. (ICBME) : The Bio-Era: New Challenges, New Frontiers*, pages 4–7, Singapore, December 2002.

[484] J.G. Linvill and J.C. Bliss. A direct translation reading aid for the blind. *Proc. IEEE*, 54(1):40–51, January 1966.

[485] J.S. Brugler, J.D. Meindl, J.D. Plummer, P.J. Salsbury, and W.T. Young. Integrated electronics for a reading aid for the blind. *IEEE J. Solid-State Circuits*, SC-4:304–312, December 1969.

[486] J.D.Weiland, W. Liu, and M.S. Humayun. Retinal Prosthesis. *Annu. Rev. Biomed. Eng.*, 7:361–404, 2005.

[487] W.H Dobelle, M.G. Mladejovsky, and J.P. Girvin. Artifical vision for the blind: Electrical stimulation of visual cortex offers hope for a functional prosthesis. *Science*, 183(123):440–444, 1974.

[488] W.H. Dobelle. Artificial vision for the blind by connecting a television camera to the visual cortex. *ASAIO J.(American Soc. Artificial Internal Organs J.)*, 46:3–9, 2000.

[489] C. Veraart, M.C. Wanet-Defalque, B. Gerard, A. Vanlierde, and J. Delbeke. Pattern recognition with the Optic Nerve Visual Prosthesis. *Artif. Organs*, 11:996–1004, 2003.

[490] J. Wyatt and J.F. Rizzo. Ocular implants for the blind. *IEEE Spectrum*, 33, 1996.

[491] R. Eckmiller. Learning retinal implants with epiretinal contacts. *Ophthalmic Res.*, 29:281–289, 1997.

[492] M. Schwarz, R. Hauschild, B.J. Hosticka, J. Huppertz, T. Kneip, S. Kolnsberg, L. Ewe, and H.K. Trieu. Single-Chip CMOS Image Sensors for a Retina Implant System. *IEEE Trans. Circuits & Systems II*, 46(7):870–877, July 1999.

[493] M.S. Humayun, J.D. Weiland, G.Y. Fujii, R. Greenberg, R. Williamson, J. Little, B. Mech, V. Cimmarusti, G.V. Boeme, G. Dagnelie, and E.de Juan Jr. Visual perception in a blind subject with a chronic microelectronic retinal prosthesis. *Vision Research*, 43:2573–2581, 2003.

[494] W. Liu and M.S. Humayun. Retinal Prosthesis. In *Dig. Tech. Papers Int'l Solid-State Circuits Conf. (ISSCC)*, pages 218–219, San Francisco, CA, February 2004.

[495] J.F. Rizzo III, J. Wyatt, J. Loewenstein, S. Kelly, and D. Shire. Methods and Perceptual Thresholds for Short-Term Electrical Stimulation of Human Retina with Microelectrode Arrays. *Invest. Ophthalmology & Visual Sci.*, 44(12):5355–5361, December 2003.

[496] R. Hornig, T. Laube, P. Walter, M. Velikay-Parel, N. Bornfeld, M. Feucht, H. Akguel, G. Rössler, N. Alteheld, D. L. Notarp, J. Wyatt, and G. Richard. A method and technical equipment for an acute human trial to evaluate retinal implant technology. *J. Neural Eng.*, 2(1):S129–S134, 2005.

[497] A. Y. Chow, M. T. Pardue, V. Y. Chow, G. A. Peyman, C. Liang, J. I. Perlman, and N. S. Peachey. Implantation of silicon chip microphotodiode arrays into the cat subretinal space. *IEEE Trans. Neural Syst. Rehab. Eng.*, 9:86–95, 2001.

[498] A.Y. Chow, V.Y. Chow, K. Packo, J. Pollack, G. Peyman, and R. Schuchard. The artificial silicon retina microchip for the treatment of vision loss from retinitis pigmentosa. *Arch. Ophthalmol.*, 122(4):460–469, 2004.

[499] E. Zrenner. Will Retinal Implants Restore Vision? *Science*, 295:1022–1025, February 2002.

[500] E. Zrenner, D. Besch, K.U. Bartz-Schmidt, F. Gekeler, V.P. Gabel, C. Kuttenkeuler, H. Sachs, H. Sailer, B. Wilhelm, and R. Wilke. Subretinal Chronic Multi-Electrode Arrays Implanted in Blind Patients. *Invest. Ophthalmology & Visual Sci.*, 47:E–Abstract 1538, 2006.

[501] D. Palanker, P. Huie, A. Vankov, R. Aramant, M. Seiler, H. Fishman, M. Marmor, and M. Blumenkranz. Migration of Retinal Cells through a Perforated Membrane: Implications for a High-Resolution Prosthesis. *Invest. Ophthalmology & Visual Sci.*, 45(9):3266–3270, September 2004.

[502] D. Palanker, A. Vankov, P. Huie, and S. Baccus. Design of a high-resolution optoelectronic retinal prosthesis. *J. Neural Eng.*, 2:S105–S120, 2005.

[503] D. Palanker, P. Huie, A. Vankov, A. Asher, and S. Baccus. Towards High-Resolution Optoelectronic Retinal Prosthesis. *BIOS*, 5688A, 2005.

[504] H. Sakaguchi, T. Fujikado1, X. Fang, H. Kanda, M. Osanai, K. Nakauchi, Y. Ikuno, M. Kamei, T. Yagi, S. Nishimura, M. Ohji, T. Yagi, and Yasuo Tano. Transretinal Electrical Stimulation with a Suprachoroidal Multichannel Electrode in Rabbit Eyes. *Jpn. J. Ophthalmol.*, 48(3):256–261, 2004.

[505] K. Nakauchi, T. Fujikado, H. Kanda, T. Morimoto, J.S. Choi, Y. Ikuno, H. Sakaguchi, M. Kamei, M. Ohji, T. Yagi, S. Nishimura, H. Sawai, Y. Fukuda, and Y. Tano. Transretinal electrical stimulation by an intrascleral multichannel electrode array in rabbit eyes. *Graefe's Arch. Clin. Exp. Ophthalmol.*, 243:169–174, 2005.

[506] M. Kamei, T. Fujikado, H. Kanda, T. Morimoto, K. Nakauchi, H. Sakaguchi, Y. Ikuno, M. Ozawa, S. Kusaka, and Y. Tano. Suprachoroidal-Transretinal Stimulation (STS) Artificial Vision System for Patients with Retinitis Pigmentosa. *Invest. Ophthalmology & Visual Sci.*, 47:E–Abstract 1537, 2006.

[507] A. Uehara, K. Kagawa, T. Tokuda, J. Ohta, and M. Nunoshita. A CMOS retinal prosthesis with on-chip electrode impedance measurement. *Electron. Lett.*, 40(10):582–583, March 2004.

[508] Y.-L. Pan, T. Tokuda, A. Uehara, K. Kagawa, J. Ohta, and M. Nunoshita. A Flexible and Extendible Neural Stimulation Device with Distributed Multichip Architecture for Retinal Prosthesis. *Jpn. J. Appl. Phys.*, 44(4B):2099–2103, April 2005.

[509] A. Uehara, Y.-L. Pan, K. Kagawa, T. Tokuda, J. Ohta, and M. Nunoshita. Micro-sized photo detecting stimulator array for retinal prosthesis by distributed sensor network approach. *Sensors & Actuators A*, 120(1):78–87, May 2005.

[510] T. Tokuda, Y.-L. Pan, A. Uehara, K. Kagawa, M. Nunoshita, and J. Ohta. Flexible and extendible neural interface device based on cooperative multichip CMOS LSI architecture. *Sensors & Actuators A*, 122(1):88–98, July 2005.

[511] T. Tokuda, S. Sugitani, M. Taniyama, A. Uehara, Y. Terasawa, K. Kagawa, M. Nunoshita, Y. Tano, and J. Ohta. Fabrication and validation of a multichip neural stimulator for *in vivo experiments toward retinal prosthesis*. *Jpn. J. Appl. Phys.*, 46(4B):2792–2798, April 2007.

[512] K. Motonomai, T. Watanabe, J. Deguchi, T. Fukushima, H. Tomita, E. Sugano, M. Sato, H. Kurino, M. Tamai, and M. Koyanagi. Evaluation of Electrical Stimulus Current Applied to Retina Cells for Retinal Prosthesis. *Jpn. J. Appl. Phys.*, 45(4B):3784–3788, April 2006.

[513] T. Watanabe, K. Komiya, T. Kobayashi, R. Kobayashi, T. Fukushima, H. Tomita, E. Sugano, M. Sato, H. Kurino, T. Tanaka, M. Tamai, , and M. Koyanagi. Evaluation of Electrical Stimulus Current to Retina Cells for Retinal Prosthesis by Using Platinum-Black (Pt-b) Stimulus Electrode Array. In *Ext. Abst. Int'l Conf. Solid State Devices & Materials (SSDM)*, pages 890–891, Yokohama, Japan, 2006.

[514] A. Dollberg, H.G. Graf, B. Höfflinger, W. Nisch, J.D. Schulze Spuentrup, K. Schumacher, and E. Zrenner. A Fully Testable Retinal Implant. In *Proc. Int'l. Conf. Biomedical Eng.*, pages 255–260, Salzburg, June 2003.

[515] E. Zrenner. Subretinal chronic multi-electrode arrays implanted in blind patients. In *Abstract Book Shanghai Int'l Conf. Physiological Biophysics*, page 147, Shanghai, China, 2006.

[516] W. Liu, P. Singh, C. DeMarco, R. Bashirullah, M. Humayun, and J. Weiland. Semiconductor-based Implantable Microsystems. In W. Finn and P. LoPresti, editors, *Handbook of Neuroprosthetic Methods*, chapter 6, pages 127–161. CRC Pub. Company, Boca Raton, FL, 2002.

[517] G.F. Poggio, F. Gonzalez, and F. Krause. Stereoscopic mechanisms in monkey visual cortex: binocular correlation and disparity selectivity. *J. Neurosci.*, 8(12):4531–4550, December 1988.

[518] D.A. Atchison and G. Smith. *Opitcs of the Human Eye.* Butterworth-Heinemann, Oxford, UK, 2000.

[519] B. Sakmann and O.D. Creutzfeldt. Scotopic and mesopic light adaptation in the cat's retina. *Pflögers Arch.*, 313(2):168–185, June 1969.

[520] J.M. Valeton and D. van Norren. Light adaptation of primate cones: An analysis based on extracellular data. *Vision Research*, 23(12):1539–1547, December 1983.

[521] V.C. Smith and J. Pokorny. Spectral sensitivity of the foveal cone photopigments between 400 and 500 nm. *Vision Research*, 15(2):161–171, February 1975.

[522] A. Roorda and D.R. Williams. The arrangement of the three cones classes in the living human eye. *Nature*, 397:520–522, February 1999.

[523] R.M. Swanson and J.D. Meindl. Ion-implanted complementary MOS transistors in low-voltage circuits. *IEEE J. Solid-State Circuits*, SC-7(2):146–153, April 1972.

[524] Y. Taur and T. Ning. *Fundamental of Modern VLSI Devices.* Cambridge University Press, Cambridge, UK, 1988.

[525] N. Egami. Optical image size. *J. Inst. Image Information & Television Eng.*, 56(10):1575–1576, November 2002.

Index